ENVIRONMENTAL DEGRADATION

ENVIRONMENTAL DEGRADATION

Chandra Mani

Oxford Book Company
Jaipur ■ New Delhi

Published By

Oxford Book Company

Regd. Off. "Bony Residency", Gate No. 2
Opp. Tilak Public School, Vishweswariya Nagar,
Gopalpura Road, Jaipur - 302018, India
Ph: 0141 - 2761280, Telefax: 2761381
email - oxfordbookjpr@gmail.com
website - oxfordbookcompany.com

Delhi Office: 102, 1st Floor, Satyam House,
4327/3, Ansari Road, Darya Ganj,
New Delhi - 110002, India
Ph: 011-45652440

Distributed By
KEYA INTERNATIONAL PRESS LLP
E-mail: keyainternationalpressllp@gmail.com

Published in India, 2025

ISBN : 9789380179698

Typeset by Shivangi Computers, Jaipur

Printed and bound in India by Thomson Press Ltd.,
New Delhi

Preface

The problem of degradation of environment due to excessive and reckless exploitation and unscientific management of natural resource has emerged as a challenge at the global level. The soil, water and air serve as major abiotic environmental sinks for the pesticides, industrial effluents, urban wastes and radioactive wastes. The pollution of these abiotic components, ultimately, endangers the very survival of modern civilization. Abatement of environmental degradation and protection of natural resources in an absolute imperative for attaining the goal of sustainable development.

Pressure of growing human and livestock population on our forests/ pastures and man's greed for more industrialisation for material prosperity, have resulted in relentless exploitation of natural resources and degradation of forests. Pollution is yet another threat to the environment. The increased traffic and industrial activities cause pollution of air and water. These days, we breathe in more carbon monoxide, carbon dioxide and harmful materials than permissible limits.

The present book provides a comprehensive and authoritative account from viewpoints of ecology, toxicology, chemistry, botany, biochemistry, soil science and biotechnology as well as impact of environmental degradation on man and biosphere. It contains chapters on various aspects of environmental degradation, technologies for resource conservation and equipments and tools for restoration of degraded ecosystems.

Chandra Mani

Contents

Contents

Chapter 1

Environmental Degradation

Environmental degradation is a process through which the natural environment is compromised in some way, reducing biological diversity and the general health of the environment. This process can be entirely natural in origin, or it can be accelerated or caused by human activities. Many international organizations recognize environmental degradation as one of the major threats facing the planet, since humans have only been given one Earth to work with, and if the environment becomes irreparably compromised, it could mean the end of human existence.

There are a number of ways in which environmental degradation can work. In a classic case, resources simply become depleted. Air, water, and soil are all resources which are vulnerable to depletion through overuse, as are natural resources like minerals and oil deposits. Habitat pressures which force animals into a small area can also contribute to resource depletion, as the animals consume a high volume of material in a small area.

Pollution is another cause of environmental degradation. When the environment becomes polluted, it means that toxic substances have rendered it unhealthy. Pollution can come from a variety of sources, including vehicle emissions, agricultural runoff, accidental chemical release from factories, and poorly-managed harvesting of natural resources. In some cases, pollution may be reversible with costly environmental remediation measures, and in other instances, it may take decades or even centuries for the environment to cope with the pollution.

Simple damage is also a common issue. Clearcutting, unsustainable development, and erosion are all forms of environmental damage. If the damage is extensive, the environment may not be able to reach a state of balance on its own, and the problem could become compounded. Erosion as a result of bad agricultural practices, for example, can strip the earth of its valuable topsoil, leaving coarse, useless soils behind.

This infamously occurred in North America during the Dust Bowl of the 1930s, in which drought, poor farming practices, and severe weather led to a widespread stripping of fertile topsoil from farmlands.

A number of social and legal issues are involved in environmental degradation, ranging from the need to provide living space for humans to questions about who is responsible for environmental cleanup. For example, if a company acquires a company which released toxic chemicals into the environment in an era when this practice was commonplace, it may argue that it has no legal obligation to clean up the chemicals, although it may be obliged to do so under ethical principles.

Environmental degradation is the deterioration of the environment through depletion of resources such as air, water and soil; the destruction of ecosystems and the extinction of wildlife.

Environmental degradation is one of the ten threats officially cautioned by the High Level Threat Panel of the United Nations. The World Resources Institute (WRI), UNEP (the United Nations Environment Programme), UNDP (the United Nations Development Programme) and the World Bank have made public an important report on health and the environment worldwide on May 1, 1998.

Environmental degradation is of many types. When natural habitats are destroyed or natural resources are depleted, environment is degraded.

Environmental Change and Human Health, a special section of World Resources 1998-99 in this report describes how preventable illnesses and premature deaths are still occurring in very large numbers. If vast improvements are made in human health, millions of people will be living longer, healthier lives than ever before.

In these poorest regions of the world an estimated one in five children will not live to see their fifth birthday, primarily because of environment-related diseases. Eleven million children die worldwide annually, equal to the combined populations of Norway and Switzerland, and mostly due to malaria, acute respiratory infections or diarrhoea — illnesses that are largely preventable.

LAND DEGRADATION

Land degradation is a complex ensemble of surface processes (e. g. wind erosion, water erosion, soil compaction, salinisation, and soil water-logging). These can ultimately lead to "desertification". As the increasing world population places more demands on land for food production etc. , many marginal arid and semiarid lands will be at risk of degradation.

The need to maintain sustainable use of these lands requires that they be monitored for the onset of land degradation so that the problem may be addressed in its early stages. Monitoring will also be required to assess the effectiveness of measures to control land degradation.

Fig. View of Land Degradation

Land is the most important basic natural resource. It is a dynamic and complex combination of geology, topography, hydrology, soil and flora and fauna and influences every sphere of human activity. Different sectors including agriculture, industries, infrastructure, and power projects put forth competing demand for land. Subsistence farming practices, accelerated soil and water erosion, erratic rainfall, increasing population, and high density of livestock population have all contributed to unsustainable land use that has lead to degradation of this valuable resource, in Karnataka.

Land is classified into two categories, arable and non arable. Non arable land comprises of area under forests, permanent pasture land, current fallow, cultivable waste land and land put to non agricultural use. Arable land includes area sown with crops (net sown area), area sown more than once and gross sown area. The extent of non arable land is 60. 50 lakh hectares.

Out of the total non-arable land in the state, 9. 67 per cent are rock lands. High proportion of rock lands occur in dry and coastal zones (about 15 per cent each). Non-arable lands are strongly gravelly in about 79 per cent in the state and a very high proportion (99. 91 per cent) is found in dry and transition zones (93. 12 per cent).

About 27 per cent of these lands have high slopes, high proportion of slopy land occurs in hill zone (50. 85 per cent) and coastal zone (35. 32 per cent). Soil with less than 25 cm depth occurs in dry (39. 6 per cent) and transition zones (47. 60 per cent). Erosion is a problem associated with nonarable land and 54. 51 per cent of the non-arable lands are severely eroded. Severe erosion of non-arable lands is a major problem particularly in dry zones (73. 55 per cent) and transition zones (97. 05 per cent).

The land and forest degradation has caused severe soil erosion in the maidan areas of Karnataka. The observed average rate of sedimentation is ranging between 2. 19 to 23. 59 hectare-meter/100 square kilometers, where as the threshold level is between 0. 29 to 4. 29 hectares. Most of the tanks have been silted up to more than 30 per cent of their capacities reducing their command area by 35 per cent.

The rate of silt deposition in irrigated tanks is estimated at 8. 51 hectare-metre/100 square kilometers/year against the assumed siltation of 3. 02 hectare-metres/100 square kilometers/year. As per the estimates of the State Government about half of arable land in the State needs protection.

Out of 125. 85 lakh hectares, 68 lakh hectares (57 per cent) needs soil conservation. The salinization has become acute problem in the command areas of the State. It is reported that nearly 10 per cent of the total irrigated area in the State is subjected to water logging, salinity and alkalinity.

Soil and water erosion has caused soil fertility loss, thus reducing its productive capacity. The soil structure, texture and moisture holding capacity are also affected due to soil erosion. Excessive chemical usage for agriculture is a problem in the cotton growing areas of the state. The average fertiliser used for growing paddy and sugarcane usually ranges between 80-150 kliograms/hectare. The arid regions of the state where rainfall is low are vulnerable to soil erosion.

Table. Presentation of Hotspots

Issues	Districts
Excessive chemical use	Shimoga, Mandya, Bellary, Raichur
Excessive Pesticide use	Gulbarga, Bijapur, Raichur
Soil erosion	All northeastern plain Districts
Water erosion	Coastal and Western Ghat districts

THE UNDERLYING CAUSES OF ENVIRONMENTAL DEGRADATION

Environmental degradation is a result of the dynamic inter play of socio-economic, institutional and technological activities. Environmental changes may be driven by many factors including economic growth, population growth, urbanization, intensification of agriculture, rising energy use and transportation. Poverty still remains a problem at the root of several environmental problems.

SOCIAL FACTORS

Population

Population is an important source of development, yet it is a major source of environmental degradation when it exceeds the threshhold limits of the support systems. Unless the relationship between the multiplying population and the life support system can be stabilized, development programmes, howsoever, innovative are not likely to yield desired results.

Population impacts on the environment primarily through the use of natural resources and production of wastes and is associated with environmental stresses like loss of biodiversity, air and water pollution and increased pressure on arable land.

India supports 17 per cent of the world population on just 2. 4 per cent of world land area. It's current rate of population growth at 1. 85 per cent continues to pose a persistent population challenge. In view of the linkages between population and environment, a vigorous drive for population control need hardly be over emphasised.

Poverty

Poverty is said to be both cause and effect of environmental degradation. The circular link between poverty and environment is an extremely complex phenomenon. Inequality may foster unsustainability because the poor, who rely on natural resources more than the rich, deplete natural resources faster as they have no real prospects of gaining access to other types of resources.

Moreover, degraded environment can accelerate the process of impoverishment, again because the poor depend directly on natural assets. Although there has been a significant drop in the poverty ratio in the country from 55 per cent in 1973 to 36 per cent in 1993-94, the absolute number of poor have, however, remained constant at around 320 million over the years. An acceleration in poverty alleviation is imperative to break this link between poverty and the environment.

Urbanisation

Lack of opportunities for gainful employment in villages and the ecological stresses is leading to an ever increasing movement of poor families to towns. Mega cities are emerging and urban slums are expanding. There has been an eightfold increase in urban population over 1901-1991. During the past two decades of 1971-91, India's urban population has doubled from 109 million to 218 million and is estimated to reach 300 million by 2000 AD.

Such rapid and unplanned expansion of cities has resulted in degradation of urban environment. It has widened the gap between demand and supply of infrastructural services such as energy, housing, transport, communication, education, water supply and sewerage and recreational amenities, thus depleting the precious environmental resource base of the cities.

The result is the growing trend in deterioration of air and water quality, generation of wastes, the proliferation of slums and undesirable land use changes, all of which contribute to urban poverty.

ECONOMIC FACTORS

To a large extent, environmental degradation is the result of market failure, that is, the non existent or poorly functioning markets for environmental goods and services. In this context, environmental degradation is a particular case of consumption or production externalities reflected by divergence between private and social costs (or benefits).

Lack of well defined property rights may be one of the reasons for such market failure. On the other hand, Market distortions created by price controls and subsidies may aggravate the achievement of environmental objectives.

The level and pattern of economic development also affect the nature of environmental problems. India's development objectives have consistently emphasised the promotion of policies and programmes for economic growth and social welfare. Between 1994-95 and 1997-98, the Indian economy has grown a little over 7 per cent per annum: the growth of industrial production and manufacturing averaging higher at 8. 4 per cent and 8. 9 per cent.

ENVIRONMENTAL DEGRADATION AND ITS PROTECTION

"O Zone" — A sound, first time heard by the homo-sapiens of this earth in the early Eighties. Some were confused and some were inquisitive. What is it? "O Zone layer" discovered by the scientists and sent an alarm signal all over the world about the impending danger and environmental disaster of our dear earth.

"O Zone Layer" created by the emission of poisonous gases from the industrial hubs in the upper sphere of the atmosphere and through the said zones, the deadly Ultra Violet Ray find its easy access to the bed of the earth. As a result it creates deadly health hazards to all the living beings of this earth. This has awakened the mind of the people to do something to minimize and eradicate environmental disaster from this earth.

Thereafter "Kyoto Protocol" was signed by all the countries sans United States of America which ranked first as the industrial polluting country in the world and the said protocol has become effective with effect from 16th February'2005 and all signatory country agreed to abide by the norms to minimize and eradicate industrial pollution.

"Environment" and its degradation and preservation have been a burning topic at present all over the world. In the last two decades, an increased concern for the purity of environment has been by many countries and witnessed a rapid and extensive legislative activity to check whatever that polluted the environment. In today's scenario international community is very concerned and vigilant on its preservation all over the world.

The word "Environment" is of broad spectrum which brings within its ambit, Hygienic atmosphere and Ecological balance. It is therefore not only the duty of the State but also the duty of every citizen to maintain hygienic environment. Many are the noteworthy judicial decisions that touched upon the various aspects environmental pollution and degradation and all these development have found an echo in India too.

Article 21 of the Constitution of India envisages right to life as a fundamental right and therefore enjoyment of life and its attainment including the right to life with human dignity encompasses within its ambits.

The protection and preservation of environment, ecological balance free from pollution of air and water, sanitation without which life can not be enjoyed.

Pollution of environment, ecological, air, water regarded as violation of Article 21. Therefore hygienic environment is an integral facets of right to healthy life. Recently, Supreme Court had made many land mark judgments for the preservation of environment, ecology, wildlife, forests,etc.

Some of these are Taj Mahal case checkmating the pollution, forest of North-East case relating to ban on felling of trees, Church Gate case restricting noise pollution created by religious prayer through audio amplifier system, etc. Therefore enforcement of Article 21 of the Constitution of India has become an effective tool for the preservation of environment and ecology.

For the preservation of environment, the Central Government and State Governments had enacted many statutes like, The Environment

Protection Act'1986, Wildlife Protection Act'1972, Madras Town Nuisance Act'1889, etc. As a part to educate the people, the Central Government had launched National Environment Awareness Campaign through the Ministry of Environment and Forest every year since 1986 with the objective of creating environmental awareness at the national level.

Through the evolution of earth, nature was endowed with an environment which was conductive to the production of life and ultimately evolution of human life. Since that time environment consists of two components –

- *Physical environment*: Includes chemical and geographical, etc.
- *Biological environment*: Consisting of plants and animals which are known as biosphere.

Broadly, environment may be divided into four major elements--

- Land,
- Water,
- Air
- Living organism.

Environmental Pollution occurs in various ways and pollution of air, water, land is alarmingly moving northwards. Pollution is categorized into the following division –

POLLUTION

Radiation Pollution: For whatever purposes nuclear energy is used – be it for peaceful purpose or nuclear warfare – the physical and bio environment cannot get way with its hazards. These are the delayed effect of fall out and increasingly important hazards of peacetime user of radiation.

Air Pollution: All most all chemicals released into air eventually find their way to atmosphere. Chemicals such as pesticides are often sprayed as mists which may drift for several kilometers from their point of application.

Pollutants like asbestos fibres, which can cause virulent lung cancer which can occur many years after the actual exposure. Similarly, nitrogen oxides exhausted from the car are component of automobile smog and nickel and chromium containing dusts which can initiate lung cancer.

Water Pollution: Pollution of fresh water is one of the most serious environmental problems for the world as a whole. Industrial process requires huge quantity of water but at the same time dilute waste products to the river. Serious sources of water pollution is the leaching of sulphide from mine working and dump and became toxic which flows down to river causing serious contamination of water and health hazards.

Soil Pollution: Soil in a forest is liable to get eroded quickly through the rain or wind unless it is adequately covered by the vegetation. Researchers initially blamed the climatic condition, today suspicion focused on environmental pollution, particularly the acid rain caused auto and industrial emissions. Acid rain prevents micro organism in soil from converting organic debris into fertilizer.

Noise Pollution: Noise usually defined as unwanted sound or sound without value. Noise as an environmental pollutant becomes an object of serious social concern. Yet noise is a persistent pollutant in many practical situations where noise sources cannot be suppressed and degrades the quality of life.

ECOSYSTEMS

Environmentalists are worried over the ill effects of increased burning of fossil fuels and of late fire woods without equivalent plantation. Such burning emits carbon dioxide and changes the temperature pattern of the atmosphere by diffusion of heat into space, a phenomenon known as "Green House effect".

Realizing the potential threat, " CHIPKO MOVEMENT" was launched in the year 1970 at Gopeswar (Garwal district of Uttaranhal). Scores of neighbourly villages were divested by a flash flood in Alakananda river.

There was clear evidence that the flood had been caused by man made erosion due to large scale felling of trees in this Himalayan region. When strong shower falls on the slope of the hill, they wash away the loose soil cover and only trees can hold the soil cover. Sri C. P. Bhatt, the leader of the "Chipko Movement", at that point of time said that "We will embrace (chipko) the trees.

If they want to cut them, their axes will first have to fall on us". This was told to the forest contractors in March'1973 who came to Reni village near Tibetan border to cut the trees. Thus these tree hugging (Chipko) programme spread spontaneously and several eco-development camps were organized.

WILDLIFE

A major factor responsible for the decline of wildlife all over the world is trade and commerce. This aroused such concern that an international treaty was drawn up in 1973 to protect wildlife against such over exploitation and prevent international trade from threatening endangered species from extinction.

The aforesaid treaty known as CITES, The Convention on International Trade in Endangered Species of Wild Flora and Fauna, and the said treaty came into force from 1st July'1975 and ratified by 120 countries.

The notion that, the public has a right to expect certain lands and natural areas to retain their natural characteristics finding its way into the law of the land. Ancient Roman Empire developed a legal theory known as the "Doctrine of the Public Trust".

The Public Trust Doctrine primarily rests on the principle that certain resources like, air,sea, waters and forests have such a great importance to the people as a whole that it would be wholly unjustified to make them a subject of private ownership and such resources being gift of nature should be made freely available to everyone irrespective of the status in life.

The public trust doctrine under the English Common Law extended only to certain traditional uses like such as navigation, commerce and fishing. The American Court in recent cases expanded the concept of The public trust doctrine and the observation of Supreme Court of United States of America in Monolake case clearly show the judicial concern in protecting all categorically important land, fresh water, wet lands

The observation therein to the effect that the protection of ecological value is among the purpose of public trust and may give rise to an argument that the ecology and the environmental protection is a relevant factor and the court in United States of America finally began to adopt this reasoning and doctrine. English Common Law includes Public Trust Doctrine as part of its juris product. Environmental Law has now become a specialized field.

In the decision which was taken on the United Nations Conference on Environment and Development held at Rio de Janeiro in June'1992 in which India also participated and the States were called upon to develop natural laws regarding liability and compensation for the victims of pollution and other environmental damages. With increasing threats to the environmental degradation taking place in different part of the country, it may not be possible for any single authority to effectively control the same.

"Sustainable development" has come to be accepted as viable concept to eradicate poverty and improve the quality of human life. Essential feature of "Sustainable development" are as follows –

- *Precautionary principle*: This principle in the context of municipal law means environmental measures by the State Government and the statutory authorities must anticipate, prevent and attack the causes of environmental degradation. The onus of proof is on the actor or the developer/ industrialists to show that his action is environmentally benign. Precautionary principle has been accepted as part of the law of the land and Article 21,47, 48-A and 51-A(g) of the Constitution of India gave clear mandate to the State to protect and improve the environment and to safeguard the forest and wildlife of the country.

- *The polluter Pay Principle:* This principle is a sound principle and interpreted by the Supreme Court that absolute liability to harm to the environment extends not only to compensate the victims but also cost of restoring the environmental degradation. As such the polluter is liable to pay the cost to the individual sufferers as well as the cost of reversing the damaged ecology.

LEGISLATIONS

Various laws has been framed in India for the protection of environment and some of these are cited below-

- Section 268 to 290 of Indian Penal Code deals with public nuisances. Public nuisance means pollution of air, water, blasting, excessive smoke, filth and other polluting activities.
- Section 133 and 143 of Code of Criminal Procedure Code and Section 91 of Code of Civil Procedure envisages that a person may approach a Magistrate and District Judge respectively by filing a complain or petition about the public nuisance.
- Under Law of Torts, special damage can be claimed from nuisance maker/violator of environment.
- The Water (Prevention and Control of Pollution) Act'1974,
- The Environment Protection Act'1986,
- Wildlife (Protection) Act'1972,
- The Air (Prevention and Control of Pollution) Act'1981,
- The Prevention of Cruelty of Animals Act'1960,
- The National Environment Tribunal Act'1995.

Problem of pollution is the outcome of urbanization, overpopulation and industrialization. In modern times, therefore, it needs more effective legal opinions to counter the above. Accordingly Indian parliament passed The Environment Protection Act'1986 to safeguard the environmental degradation. The Indian Penal Code has few provisions on the subject, but they are ineffective when faced with the problems of an industrialized society.

The first problem to attract the attention of certain state legislation

in India was water pollution. But it was only in 1974 that a Central Act was enacted on the subject to be followed by The Water (Prevention and Control of Pollution) Cess Act'1977 and thereafter most drastic law had been enacted as Environment (Protection) Act'1986. India first got the taste of environmental disaster by two catastrophes that befell India – the Bhopal disaster in 1984 and Sri Ram Fertilizer Plant leak in 1985.

The Bhopal Gas Leak Disaster (Processing of claims) Act'1985 gave the Central Government the sole authority to represent (in litigation) the victims of Bhopal for compensation claims against the Union Carbide Company.

Sri Ram gas Leak did not generate legislative activity but prompted Mr. M. C. Mehta and certain other organizations to start to use the effective legal tool in the form of Public Interest Litigation (PIL). This gave the Supreme Court an opportunity to enunciate certain important doctrine on tort law, corporate law (particularly the civil liability of directors for wrongs committed by the corporate body).

Recently, Supreme Court has broadly and liberally interpreted the Article 21 and transgressed into the area of protection of environment and held that the protection of environment and citizen's right to live in eco-friendly atmosphere interpreted as the basic right guaranteed under Article 21.

Recently to mitigate the needs of environment related litigation, "Green Benches" had been constituted in many High Courts in the Country. Some of the following decisions of Supreme Court of India has a great ramification towards the protection and safeguarding the environment and maintain the ecological balance.

COMBATING ENVIRONMENTAL DEGRADATION

The landmark report of the World Commission on Environment and Development, entitled "Our Common Future", warned that unless we change many of our lifestyle patterns, the world will face unacceptable levels of environmental damage and human suffering.

The Commission, echoing the urgent need for tailoring the pace and the pattern of global economic growth to the planet's carrying capacity, said that: "Humanity has the ability to make development sustainable and to ensure that it meets the needs of the present without compromising the ability of future generations to meet their own needs. "

In the final analysis, the environmental crisis affects everyone on the planet, but the degree to which the inhabitants of different parts of the world contribute to this crisis depends on the level of their economic development and their consumption patterns. As much as 70% of the world's consumption of fossil fuel and 85% of chemical products is attributable to 25% of the world's population.

Water consumption is also unevenly distributed. The per caput water consumption in the United States is about 2 300 m³ per annum, as compared to 1 500 m³ for the Canadians and 225 m³ for the British.

The average per caput consumption of water in developing countries ranges between 20 to 40 m³. The consumption patterns for forest products and many other commodities have the same direct inverse proportion to the size of population of the top 20% of the richest societies. This profligate demand puts excessive pressure on both national and global natural resources. The rest of the world, comprising 80% of its population with a share of less than 20% of global income, has a far more modest consumption level.

While international environmental concerns are often expressed in broad terms such a desertification or climatic change, the environmental problems of concern to vulnerable groups in marginal areas are generally quite localized in nature, revolving around immediate issues, such as the degradation of a particular rangeland or soil erosion on farmland or the progressive shortening of fallow. These affect the poor because they are directly related to household food security. Degradation of the resource base generally translates into decreases in production or income and thus in the availability of food.

Declining soil fertility leads to lower crop yields while rangeland depletion reduces offtake, and any deterioration in water quality adversely affects the fish catch. Degradation of common property resources pulls labour away from directly productive activities towards gathering - simply collecting non-wood and minor forest products - and probably diminishes opportunities for deriving income from this source.

Linkages with food security can also be less direct. Shortages of biomass may result in a transition to lower-nutrition foods that require less fuel for cooking. In addition, recurrent drought or natural calamities also directly result in progressive loss of food security prospects.

In their quest for food security, the rural poor have sometimes little choice but to overuse the limited resources available to them. The resulting environmental degradation imposes further constraints on their livelihood in what has been called a "downward spiral" or "vicious circle". They are often forced to make trade-offs between immediate household food requirements and environmental sustainability both in production and consumption. Their negligible man-made capital assets, ill-defined or non-existent property rights, limited access to financial services and other markets, inadequate safety nets in time of stress or disaster, and lack of participation in decision-making can result in their adopting "short time horizons", which favour immediate imperatives over longer-term objectives.

This can result in coping strategies that rely on the drawing down of the capital available to them — mainly in the form of natural resources. It also makes them more vulnerable to environmental degradation, including degradation wrought by others than the poor themselves.

The poor may be both agents and victims of environmental degradation, especially in marginal areas, where the resource base is ill-suited to agriculture. But it cannot be assumed that the poor have an intrinsic propensity to degrade environmental resources.

On the contrary, many poor traditional communities demonstrate an admirable environmental ethic and have developed complex resource management regimes. There is little evidence that the rural poor, when offered an appropriate environment - including secure tenure and access to markets- pursue resource-degrading strategies.

Thus, while poverty may be an underlying cause of environmental degradation, it is more accurately seen as a proximate cause influenced by a complex of policy and institutional factors. The very same processes that lead to and perpetuate poverty constrain the poor in their decision- making with regard to natural resource management. Affluence and poverty affect the environment in different ways: poverty eradication would not erase environmental degradation but change the nature of environmental problems facing society.

POVERTY IN FRAGILE ECOSYSTEMS

Absolute poverty has been on the retreat in most high-potential areas in developing countries. The combination of more productive technologies, fertile land and water, and high levels of development and public investment have raised incomes significantly for people living in these areas.

While this development has not always been equitable - or sustainable, the most important disparities are not between rich and poor people within high- potential areas, but rather between high-potential high-investment areas and fragile ecosystems.

In the latter areas, politically marginal indigenous populations have been neglected and have been joined by new groups displaced from more fertile areas through a variety of processes. These processes, although varying across countries and regions, include expropriation, demographic pressures, land fragmentation, privatization of common property lands, and consolidation and expansion of the commercial sector combined with reduced demand for labour due to mechanization.

While the challenge for poverty alleviation in high-potential areas remains considerable, the prognosis is not grim provided agricultural intensification proceeds without environmental destruction.

On the other hand, for the 60% of poor populations who are found in fragile ecosystems and mainly remote and ecologically vulnerable rural areas, the challenge of environmentally sustainable poverty alleviation is immense. It has been estimated that 80% of poor people in Latin America live in such areas, 60% in Africa and 50% in Asia.

Reliance on the currently prevailing patterns of growth will postpone the resolution of poverty in marginal areas, with severe implications not only for the people affected but also for the environment.

The immediate-to-medium-term prospects for the rural poor to abandon these areas for other sectors of the economy, as was the case in Europe in the last century, are not promising. As a result, fragile ecosystems are rapidly becoming ghettos of poverty and environmental degradation.

The need for urgent action can be recognized in relation to the following characteristics of these regions:

- They constitute a significant part of the world's land resources. Forty per cent of the earth's land surface is considered dryland, of which approximately 70% is already degraded or subject to heavy degradation. On the other hand, hilly and mountainous regions cover about 21% of the earth land mass and, although not so extensive as dry lands, they exert a far-reaching influence on other areas, primarily through watershed functions.
- The role of both ecosystems in terms of human habitat is also significant: approximately 900 million of the world's population are subsisting in dry zones. Although only about 10% of the world population live in mountain areas, a much larger per centage (about 40%) occupies the watersheds below. It is safe to assume that the future of mountain ecosystems affects the life of half of the world's population. From the Andes to the Himalayas, and from South East Asia to East and Central Africa a serious ecological deterioration caused

by overgrazing, deforestation and excessive cultivation threatens the livelihood of these populations.

- Mountains are important sources of water, energy, minerals, agricultural products and a major reserve for the world's biodiversity. Similarly, dry zones are rich in biodiversity, hosting many endangered species. Moreover, crops, grasses, trees, and livestock species, that form the core of survival in drought prone regions, exist in these regions only.
- A high proportion of the absolute poor in ecologically fragile areas are indigenous peoples, estimated at some 300 million worldwide. They depend on renewable resources to maintain their well-being. This has led to the development of livelihood systems which are well-adapted to the harsh conditions in which they lived. Their holistic, traditional knowledge of their natural resources and environment constitutes a rich human heritage. However, their traditional ways of life are now being threatened, disturbing the delicate balance of natural resource use. Nevertheless, viable technology and institutional arrangements for resource conservation in these areas could be built upon indigenous knowledge; and similarly effective disaster prevention policies can benefit from coping strategies developed by the local population.
- Rural women play a key role in on- and off-farm activities in the developing countries. This is particularly true in the case of the ecologically fragile areas. With the growing male out-migration from marginal areas, the number of women headed households in these areas is increasing. Women are becoming more and more responsible for the day to day survival of the family. Women tend to be more vulnerable than men to the effects of environmental degradation because they are often involved in harvesting common property resources such as wood and water. Since women usually make a greater contribution to household food security than men, a decline in women's access to resources may have a significant impact on household consumption. Environmental degradation implies further burdens and responsibilities which are not compensated for by increased decision-making power.
- Degradation of land and loss of its vegetative cover also have consequences at the global level, primarily because of its influence on carbon exchange, but also in terms of loss of biodiversity. The large amount of carbon stored in the vegetation of the dry zones, for example, averaging about 30

tonnes per hectare, decreases when the vegetation is depleted or disappears. Carbon-rich soils, frequently found in dry zones, store a substantial amount of this element (nearly half the total quantity of carbon is stored in the organic matter in the soil, much more than is found in the world's vegetation). The destruction of these soils has a very powerful effect on the carbon cycle and boosts the greenhouse effect as a result of the release of carbon.

TOWARDS ACTION

Over the past two decades, environmental degradation, including land degradation has continued to worsen exacerbating further poverty and food insecurity. Conversely, awareness of the importance of the environment and its conservation has increased.

There has been a transformation in people's perception of the poverty problem in developing countries. If one accepts that hard core rural poverty is increasingly a phenomenon associated with marginal lands, then new strategies are required that integrate poverty alleviation and environmental management.

Until recently, the international community and national governments have tended not to appreciate the need for integrated rural poverty alleviation and environmental management programmes in marginal areas.

There were a number of promising initiatives in this field, usually undertaken by NGOs and community- based organizations, but they were usually small and very localized. At the same time, in many regions, rural people's perception of their environment and the priority they give to a better relationship with it have changed. Increasingly, rural people are realizing that:

- The fragile environment on which they depend for their survival is being neglected or over- exploited, and it is now necessary to rehabilitate it and manage it sustainably;
- The environment belongs primarily to them, and they must take the responsibility for the land and organize themselves in groups, cooperatives, village development associations and other local association to defend it.

UNCED's Agenda 21, the global action programme for sustainable development, is perhaps the first expression of international commitment to addressing the poverty- environment nexus. "Combating poverty" called for specific long-term strategies that integrate poverty eradication and sustainable management of the environment. In the follow-up to UNCED, promising initiatives have emerged for these thematic areas. For drylands, the United Nations Convention to Combat Desertification in Those Countries Experiencing Serious Drought and/or Desertification, Particularly in Africa (CCD) provides a framework for concrete action at the local level. For mountainous areas, efforts are currently under way to develop the basis for an action plan for sustainable mountain development, known as the "Mountain Agenda".

The Agenda involves the establishment of a network on sustainable mountain development consisting of United Nations agencies, NGOs and intergovernmental institutions. A set of action proposals has been developed by those involved in promoting sustainable mountain development. In recognition of the need to give prominence to the "Mountain Agenda" on the international and national lists of priorities, a global Inter-Governmental Organizations (IGO)/NGO Conference, as well as regional inter-governmental consultations are being convened.

The main proposals for action that are emerging, identified through a broad participatory process involving the major NGOs, encompass five specific areas of focus: poverty eradication; the strengthening of a global information network and database; strengthening country capacity and the generation of "National Mountain Action Programmes"; raising awareness through the preparation and organization of a World Conference on Sustainable Mountain Development in early 1997; and the formulation, negotiation and implementation of regional or sub-regional mountain conventions and possibly the development of a "Global Mountain Charter".

The Desertification Convention offers new and exciting opportunities for collective action, as well as a fertile field for testing and nurturing innovative partnerships in development cooperation for local level action. It is the first International Treaty to squarely address poverty and environmental degradation in rural areas. Unlike the other Conventions associated with Rio, the direct beneficiaries of CCD are the hundreds of millions of predominantly poor and food-insecure people who populate the drylands of the world.

It is the first Convention that casts resource users and their communities as central to the solution rather than part of the problem. At the very heart of the CCD is the concept of "Partnership". Partnership embodies the new thrust in development assistance, in which it is finally recognized that interdependence rather than dependence is the way forward. But partnerships won't work unless all partners stand to benefit. CCD tries to translate this attractive concept into more or less concrete terms.

While the underlying incentives to enter into partnership must exist, what is also needed is a favourable context to promote its emergence and functioning. In the context of CCD, the National Action Programmes - or NAPs - are the instrument for partnership.

NAPs, which are not intended as static plans but as a dynamic programming capacity, should offer a macroeconomic and institutional framework that will support local-level action. Here, more is meant than economic and fiscal policies, although these are of course extremely important.

It also means a policy orientation that actively focuses on empowerment of local actors to take advantage of new opportunities and overcome old constraints. The Convention therefore encourages devolution of decision-making from the centre to local populations and resource users. The most important reasons for this are compellingly obvious:

- Local Ownership in Decision-making - Local structures are more likely to make decisions that are relevant and suitable to local circumstances.
- Removing Bottlenecks in Information Flow and Decision-making - Decision-making for natural resource management requires prompt and relevant information.
- Improved Ability to Involve Marginalized Groups - Decentralization might allow better targeting of services and better identification of needy groups.
- Better Tailoring of Approaches to Local Conditions - Local appreciation of constraints and opportunities can only improve the quality of solutions.

An emphasis on empowerment of local populations and civil society should not be construed as a wish to actively withdraw from the sustainable development arena. Instead, it is based on a recognition that the public sector and multilateral finance can facilitate but cannot substitute for action that must come from economic agents at the local level that act individually or collectively.

What is needed now is to build an operational coalition between NGOs, CBOs as well as other institutions of civil society together with government institutions and international agencies, to form action-oriented partnerships around specific and concrete areas of intervention.

THE CHALLENGE OF FINANCING ACTION

Promoted by the world's distress over the loss of life in the Sahelian famine of early 1970, the UN Conference on Desertification adopted a plan of action to end desertification by the close of the century. The response to the plan of action was dismal and it was virtually left on the shelf.

Now with only four years left to the day when UNCOD's promises should have materialized, desertification has almost doubled, and the poor are paying the cost, with their health and lives.

The CCD diligently negotiated and enthusiastically adopted holds new promises, as the degree of awareness, globally and locally, has increased. But unlike its sister Conventions on climate change and biological diversity, the CCD does not promote establishment of a new financial mechanism. Instead it foresees the creation of a "Global Mechanism" to be housed in an existing organization to coordinate and facilitate the flow of additional funds including grants and concessional loans through both bilateral and multilateral channels.

Neither national budgets nor statistics on international financial flows to developing countries give clear figures on resources presently allocated to combat desertification. But there is little argument about the dearth of international funding for desertification control. Even resources formally provided under Global Environmental Facility (GEF) - which, by and large, precludes eligibility for desertification programmes - are judged to be inadequate. Nevertheless, financing constitutes a major pillar for the success of CCD without which it may very well face the same fate as UNCOD. Within this context, a proactive role for the Global Mechanism should be promoted.

The multi-source and multi-channel orientation of the CCD is more of a strength than a weakness. Instead of relying on one mechanism - say, the GEF - the Convention is not predicated on the availability of external grant finance earmarked for the purpose. In contrast, the Global Mechanism configuration is about improving the effectiveness and efficiency of existing flows, in addition to catalyzing and leveraging new flows and sources of finance. It encourages a greater role for domestic resource mobilization, private sector initiative, and a blending of various concessional and non-concessional external finance.

This diversity of flows and the multifaceted diverse coalition which one hopes it would represent, will in the end make the Convention and the actions it triggers more robust and sustainable.

One should work towards that coalition, by assisting to set in place policy and institutional frameworks that are favourable to private initiative, by helping governments to provide public goods, by pump-priming promising initiatives, and by assisting local populations and community organizations to interface more productively with the private sector.

FINANCING PEOPLES' PARTICIPATION

Local-level activities and creativity championed by CCD have a number of implications for the nature of resource mobilization as well as the manner through which resources are utilized. First, there is a need to step up efforts aimed at awareness- building at local level. This is a task for which NGOs and CBOs are best suited. The NGO community, and in particular the international NGOs, should give a high priority to this objective when mobilizing resources for CCD as stipulated in the Convention.

Second, CCD calls upon Parties to promote a National Desertification Fund (NDF) and similar mechanisms for directing funds to the local level. Such mechanisms should be run on the basis of a participatory governance involving local communities and their partners in the NGO community.

NDF should also be flexible and simple in design. To preserve the confidence of both donors and local populations, it is imperative to ensure full transparency and effective accountability in its management. Moreover, the local populations could be true shareholders and effectively claim their share in the partnership if, in addition to the contribution from the external donors and national resources, they shoulder part of the financial burden. This could be done by mobilization and pooling of individual savings as well as through decentralization of collection and management of taxes, levies and other revenues derived from local resources.

Third, it is absolutely important that the NDF resources are to be utilized for community level investment and that they lead to the creation of durable economic assets, shared collectively. Using the proceeds of NDF for relief activities or financing individually- owned enterprises would be a costly mistake.

The former would deplete the resources of the fund without any lasting benefit, and the latter would distort the local financial market, preventing the creation of sound credit/saving structures. Such structures are equally important to facilitate investment for crop intensification or to promote economic diversification to lessen man and livestock pressure on land.

Populations in marginal areas are not doomed to despair. On the contrary, it is in these very regions that the people, forced by circumstance, manage to cope most creatively with their harsh and unpredictable environment, and to diversify their resource use strategies over space, season and sector. They capitalize as much as they can on biological diversity - most pronounced in these regions and constituting a core of their survival.

They are responsible for most appropriate technological and institutional innovations which depend minimally on costly and external inputs. This is particularly true in the conservation of rainwater, notwithstanding the saline soils common in those regions. It is also true for the institutions which developed for the collective management of very scarce common resources, such as water points, grazing land and forests.

Effective actions against poverty, household food insecurity, and environmental degradation in marginal areas require first and foremost the empowering and equipping of local communities to take up the reins of resource management. The importance of local area development and improved local governance - also covered in the other issues papers - must be emphasized.

An important factor in this context, of course, is the issue of incentive frameworks and enabling environments, with specific regard to the question of how to combine longer-term concerns for environmental rehabilitation and conservation with the pressing short-term needs of household food security. Also important are the technology and related measures to be promoted that build on traditional knowledge, such as those which will in the short term generate tangible benefits for the farmer, as outlined in the discussion paper on this topic.

Many conservation policies and strategies in the past have failed because of their top-down approach and their reliance on technologies which were irrelevant to the local circumstances. In contrast to the result of these efforts, the micro-projects implemented in many places over the past decade have made it possible to build up a store of knowledge allowing for the implementation of new approaches.

Within this context, a consensus has emerged on the importance of indigenous people's traditional knowledge and practices in the management of arid land, forest, pasture and farmland to conserve soil and moisture, and in diversifying crop and livestock production to minimize risks.

Some traditional rural communities have developed complex resource management systems that have stood the test of time, and have much to offer in addressing present-day concerns over long-term resource sustainability. Their admirable environmental ethic deserves its due place.

Asserting the importance of local knowledge calls for the empowerment of local people through their own organizations. Moreover, the considerable cultural and environmental heterogeneity of mountain areas and the scattered nature of dryland populations underline the need for decentralized local-level action toward integrated management of local areas.

This is not to suggest that local communities can be left to their own devices. There is a need for supportive and facilitating measures on the part of governments. The international community should also be aware of the global dimension of the process and the responsibility that this implies.

There is therefore a need for a coalition of actors ranging from the international to the national and the local level. This is precisely what the CCD is promoting and what an eventual Mountain Agenda might promote. In the short term, what is needed is what one might risk calling "affirmative action" in the form of finance and assistance to local communities.

The immediate challenge is to consider how ratification of the CCD can be expedited, how it can be implemented and how to secure adequate financing for local area development. The CCD also stipulates a major role for civil society organizations, foremost among them the community-based organizations - namely that they should galvanize energies and mobilize resources.

The private sector, as well as civil society at large, should also be encouraged to think beyond individual or corporate interests towards a recognition of a shared responsibility for the environment. Vigorous resource mobilization to combat desertification would stand a better change of succeeding if launched on the basis of empirically verifiable improvements.

Chapter 2

Eco Degradation

IMPROPER LAND USE CAUSING ECO DEGRADATION

Malnad area (Western Ghat Area) of Karnataka is a mosaic of reserve forests, community managed foliage Hillocks (*Soppina Bettas*), paddy fields and cash crop plantations. More than 50 per cent of this area is still under forest cover. Cultivations in this area depends on compost made out of foliage and leaf litter and herbal pesticides collected from these foliage hillocks. Paddy is the major food crop of this area, where a good number of farmers still follow traditional agricultural methods.

These paddy fields are rainfed uplands. Irrigation when necessary farmers make small check dams in the streams with bamboos, twigs and mud. In early days they were taking two crops per year but off-late, except in areas near the river basin there is only one crop per year due to shortage of water. Land preparation for starts in May, after summer showers and the nursery preparation start in June, after the onset of monsoon.

Koimatore sunna and high yielding varieties like IET are the most preferred rice varieties these days. Many traditional varieties are used here for cultivation. After the onset of monsoon, seedlings are transplanted to the fields. Harvest takes place in the month of December and average yield ranges from two to three tons /acre. In most parts of the Western Ghats, human use and "disturbance" of forests is the rule rather than exception.

After the pre-historic system of shifting cultivation gave way to more sophisticated multicropping systems of betel nut (Areca), coffee, pepper and cardamom in the last two centuries, human use of forests intensified. Extraction of dry and green leaves, in addition to that of fuelwood, fodder, small timber and non-timber forest products led to the creation of a patchwork of dense secondary regrowth, sparse tree savannas, and even "degraded" forests around cultivated areas.

Yet, the species occurring in these intensely used and managed forests (called *Soppina Betta, Jemma land, Derekinahadyas, Bane, Kann* etc. in local languages) are a subset of those occurring in the "natural" forests nearby.

These foliage hillocks as they are meant in local parlance are the "service tenures" over the Uncultivated Public Lands (UPLs) granted to various communities of the central Western Ghats of Karnataka state, India. These Common Pool Resources (CPRs) provide manure, fuel wood, fodder, medicine and some cash income to the community and timber for minor constructions. Apart from this, *Soppina Bettas* are a treasure house of medicinal plants on which the farmers living in this remote pocket of Western Ghats rely for various ailments.

These minor forests and its floral diversity play a pivotal role in determining the health and productivity of the agro ecosystems and sustain the livelihood of rural populations by raising the agricultural productivity. These community-managed forests are examples of slightly managed but socio-economically valuable, high yielding forests. *Soppinabetta* forests appear to be distinct land use systems that can neither be categorized as regular forests nor farmlands.

They are characterized by their sustainability in terms of plant genetic wealth and agricultural and community needs met by them. Many of the farmers of *malnad* area are still following the traditional cultivation methods.

The compost made of foliage and leaf litter collected from SBs mixed with cow dung is the major resource input to the farms. But recently many farmers have started using chemical fertilizers as a nutrient supplement. Traditional farmers still largely depend on botanicals collected from the home gardens and *Soppina Bettas*.

A recent study has revealed at least six plants species being used in this area as herbal sources of pesticides. Brown plant hoppers (BPH) are the major menace of paddy in Sringeri. Farmers opine that BPH appeared in their fields after they started using chemical fertilizers.

Many research studies from other parts of the world also are justifying this hypothesis. Farmers are not supportive of raising monoculture plantations of *Acacia auriculiformis* in degraded SBs as they feel it adversely affects the yield of paddy as it does not decompose easily. The farms near degraded SBs (scrubs) are better yielding than thickly wooded SBs with large trees, at the same time the degraded SBs are not enough for providing required green manure.

Farmers have either to bring leaf litter from far away or to turn to foliage crops like green manure crops like *Gliricedia* in the non-farm season. On the other hand, state forest policy during the colonial and post-independence period in the last century led to the complete replacement of "natural" or "degraded" forests with timber plantations of teak, eucalyptus and (more recently) *Acacia auriculiformis*. While foresters denigrate local forest use as "degrading" and local communities criticize forest plantations as "useless" and "monocultural", little is really known about the total biodiversity impacts of these alternative land-use systems, particularly the impacts on insect communities.

The topography of the area ranges from very steep to gently undulating with rocks, peaks and ridges outcropping. Altitude ranges from 623m above msl at the Sringeri Township through 1458 m above msl at Gangamoola of Kudremukh NP. Vegetation is mainly tropical evergreen and semi evergreen.

The hilltops of this region harbour grasslands and shola Forests (Tropical Montane forests). Vegetation of SBs falls in continuation with the less disturbed protected areas of this region, an area of near climatic climax vegetation of Western Ghats. Although regularly pressurized by farmers, the floral diversity of SBs are no less than the natural forests nor is their regeneration pattern significantly different.

GOODS AND SERVICES PROVIDED BY THE SYSTEM

The agro ecosystem comprising SBs, paddy and mixed stands of horticultural trees provide the staple food grain of paddy, spices, fruits and nuts apart from goods and services like manure, fuel wood, fodder, medicine and timber to the dependent communities. This system of three subcomponents take care of subsistence and health requirements. The system has been characterised with the region and has been depicted in many classic films and poems in regional language (Kannada).

The diversity of unique rainfed rice varieties for the hills as well as the floral diversity of SBs were conserved to an extent in some parts. While valuable organic matter flowed to the paddy lands and plantations, the organic systems of paddy cultivation in turn has helped sustain the forests of SBs as well as the soil water regime for the horticultural crops. The perennial stand of trees adds to the carbon sequestration benefits, which will be valuable for global climate stabilisation.

THREATS AND CHALLENGES

These economically efficient self-sustaining systems are now fast declining due to lack of care, improper management, over exploitation and lack of awareness. While free access to SBs is beneficial to the village communities, the same communities accelerate their depletion.

On one side due to over exploitation the *Soppina Bettas* are getting degraded and converted to open scrub forests which inturn is converted to *Acacia auricularis* plantations by the state forest department.

While on the other side both the landless as well as the land holding people try to improve their socio-economic conditions by converting *Soppina Bettas* to agricultural land and later approach the government or judiciary to regularize the illegal occupation and land conversion. Reckless commercial composting with large scale and continuous clearing of forest floors also pose threat to SBs.

Individual rights of fencing SBs: While SBs were allotted only for meeting the green manure requirement of limited areca palm areas it was supposed to follow a rotational period in between two years of litter collection and lopping.

Since the areca areas increased, and individual areca growers started enclosing the SBs, they are being exploited unsustainably. The major challenges faced by this GIAHS include conversion of paddy lands. More and more paddy turning to horticulture gradens mean need for more leaf litter and more extraction from SBs. It also means extracting red soil from SBs and other forests for filling paddy lands for conversion.

POLICY AND DEVELOPMENT RELEVANCE

The foliage hillocks are minor tenurial forest allocated to the *farmers* in the Western Ghat tract for deriving the foliage as green manure and to feed into the dung – litter composting. This tradition of organic subsistence to *Areca* is very old, and in early times, local communities governed the forest resource utilisation. Buchanan had recorded the practices of manuring soil with leaves of various wild trees. The purpose was to arrest the loss of topsoil, suppress weeds and conserve soil moisture.

Once East India Company took over forest administration of the country, *farmers* could not continue to depend on forests for litter and green manure. Upon hearing their grievances, the government in 1864 adopted the policy of allocating little forest area to the *farmers* in order to save rest of the forests in Uttara Kannada district. Cash crops like *Areca* requiring organic matter from SBs were taxed at higher rate than paddy and the farmers' privileges over the SBs were regularised in 1924.

The state initially allowed 4 to 8 times of minor forests per unit area of Areca holding to be utilised as the needed foliage support land. Later this allowance was extended to other districts of Western Ghats like Shimoga, Chikmagalur and Dakshina Kannada where Areca is grown.

Thus the policy repercussions include those on whether to keep extraction rights of SBs strictly individualistic or community managed, and how to restrict conversions of SBs and paddy lands to other degrading land uses. The development impact will be on sustaining the productivity of the staple crop of paddy in the region while assuring sustainability of multi-storeyed coffee-arecanut-pepper system.

This in turn leads to conserving the diversity of traditional upland paddies of the region as well as conserving diversity of SB forests. Community participation in monitoring and protecting the system should be catering to keep the dependence of the poor on SBs for fodder, firewood and subsistence to appropriate levels.

How to develop regulations on extractions and implement participatory methods of enforcement of these regulations are important questions to be answered by the project.

GLOBAL IMPORTANCE

The Western Ghats are an ancient mountain range along the western coast of India covered with tropical evergreen and deciduous forests, recognized as both a megadiversity region and a global biodiversity hotspot due to the vulnerability of its endemic biota to habitat loss.

Between 1920 and 1990, the rate of forest loss is estimated to have been 0. 57% annually. Special features of these moist tropical forests of this GIAHS in Western Ghats include:

- Dramatic change of vegetation mosaic across steep altitudinal and climatic gradients within a small expanse from west to east
- Gradual change of endemism from north to south along the long strip of the Ghats
- Old, well-weathered deep and nutrient rich soils supporting luxuriant vegetation of high diversity
- And efficient, tight nutrient –cycling capable of localized nutrient losses.

With 3000 years of human occupational history, there are records of people descending from the Ghats for shifting cultivation. The shifting cultivators preferred a zone of under 1000 m altitude in order to escape excessive cold and winds. This altitudinal region in the Western Ghats generally constitutes ill-drained valley bottoms, swamps and sluggish streams.

It is likely that such swamps were converted into Areca orchards and fields of summer rice. Eventually these agro ecosystems evolved along with human needs.

Such systems bear a range of socio economic functions along with conventional biological functions like productivity, nutrient cycling and population dynamics. *Soppina Bettas* are an example of such socio-economically valuable, high yielding non-equilibrium systems.

Where as farmer-managed system of SBs has dwindled in other districts for one reason or the other it has been functioning well in Chikmagalur district, which is a hilly and thinly populated area (<100 individuals/ km^{2}). Sringeri is the smallest Taluk of Chikmagalur district in Karnataka state, south India with a geographical area of 442. 32 km^2 (0. 23% of the state area). Sringeri town named after a sage who obtained penance in one of the hillocks situated here, boasts of a famous monastery established during 12th century AD.

The monastery at Sringeri was supported by most of the kings of south India and was recognised as a tiny semi-independent state with 23 villages under its control and managed by the pontiff of the monastery. This status was abolished in 1959 and Sringeri taluk was constituted with 49 villages inclusive of those under the monastery. The monastery encouraged forest conversion to arecanut cultivation and with population explosion started to enclose their SBs.

The taluk has a significant extent of forest cover (59%), 9% of the area is under arecanut palms and 2. 7% under paddy. SBs cover about 19% of the area. Sringeri has been selected for proposing as a pilot area for implementing GIAHS programme because among other areas with similar agro ecosystem the SBs here are fairly intact; awareness is high; proportion of SB to agricultural land is found to be optimum.

Activities foreseen to be included in such a pilot framework:

- Quantifying the value of ecosystem services of the GIAHS: This activity is aimed at showing the ecological relevance of the system in economic terms. It involves eco hydrological estimations of services and their monetisation. Monetary values will be the lower bound estimate of the total value of the system and will be helpful in convincing the measures taken to preserve the system.
- Establishing community monitoring of ecosystem health
- Facilitating stakeholder institutions for spreading and instilling awareness about the relevance of the system
- Holding awareness and workshops and policy dialogues on participatory and effective monitoring and enforcement.

These activities will be instrumental in reducing the threats described in the last section as the participatory approach to awareness creation and outreach will facilitate community involvement in reducing the challenges faced by the systemCo-financing potential ATREE has a strong presence in Sringeri for last 5 years.

We had a sub-project from McArthur foundation to study the impacts of landuse changes of SBs (Sharad Lele & Kamal Bawa for mapping vegetation and landuse changes; Priyadarsanan for litter insects). Indian Council of Agricultural Research (ICAR) under its NATP (National Agricultural Technology Project) scheme has given a three-year grant to study the impacts of the landuse changes on paddy cultivation.

Dept. of Science and Technology has provided with a grant to study the ants and dung beetles. These funding agencies, especially ICAR will be willing to extend their support if this GIAHS region attracts some international attention. ATREE is maintaining a field station in Sringeri. Our research staff there work in close collaboration with the farm community and help them sort out their day to day problems in cultivation.

DROUGHT PROOFING

This study on water and drought proofing was motivated by reports of out-migration and other drought related distress that poured in from all over the country in 2000, including the newly formed state of Chhattisgarh. It was intriguing, as well as challenging, to inquire into why a region like Chhattisgarh, which was always known as the rice bowl of Central India, should be reporting high and growing distress over the decades. Of course, nationwide the annual average precipitation has declined in the nineties.

This notwithstanding, rainfall data shows that most parts of Chhattisgarh have been receiving an average annual rainfall that cannot be characterized as chronic rainfall deficiency, which is a hallmark of many regions in the North, Northwest and West India. The chronic rain-shadow areas in Kawardha and Rajnandgaon are exceptions to this observation. Conceptually, drought proofing means the capacity to meet the basic material and physical needs of the local population - human and animal - in a drought period so that there is minimal distress. Drought is a problem of insufficient water supply, relative to normal demand.

Drought is defined as a temporary harmful and widespread lack of available water with respect to specific needs. The drought damage to crops is induced by the loss of water balance within the body of the plant. When effective moisture in the soil decreases to a certain degree, plant roots are hindered from absorbing moisture and the plant begins to wilt. Drought brings about disasters through damaging the moisture balance in the soil-plant atmospheric system.

Drought proofing an area then implies that the local natural and human production resource base can provide a certain desirable amount of food, fuel, fodder, drinking water and livelihood resources during a drought.

Drought proofing cannot totally protect the normal years' water supply patterns during the lean periods. In a drought year the production system operates at a lower level, irrespective of the drought-proofing efforts. Drought proofing must, therefore, aim at enhancing water availability during a drought so that the fall in production can be minimized. As a process, drought proofing is a continuing one that spans lean and normal years. The nature of drought-proofing works and activities during the two periods can be quite different. Water-use patterns in normal years can increase or decrease an area's vulnerability to droughts.

High water consumption patterns are more vulnerable to rainfall deficits, than lower ones. Excessive withdrawal or inadequate recharge in normal rainfall periods also increases the vulnerability during dry spells.

Thus, the normal year's water-use patterns form an integral part of drought proofing strategies. In normal years, land and water management must focus on enhancing the biomass on lands of marginal farmers and on landless people. During droughts, they must be targets of employment generation programmes.

During a drought year, drought-proofing activities include supply-side measures like conserving water use or bringing water, food or fodder from outside. On the demand side, it involves ensuring entitlements, i. e. , protecting the purchasing power of the local population through employment generation works, which, in turn, can be used to provide drought-proofing infrastructure. In sum, water availability, water use and entitlements are the three basic elements for comprehensive drought proofing.

Water availability has to be assessed and developed, water use has to be monitored and conserved, and entitlements have to be ensured for the vulnerable sections of the population, at the appropriate time. Drought proofing was enunciated as a policy goal for the first time in 1987, although scarcity management has been an important part of the administrative and policy agenda of independent India.

The early policy responses of gratuitous relief and emergency disaster management have given way to a more comprehensive and pre-emptive policy for protection against meteorological droughts. This shift from symptomatic fire fighting to protective and more proactive prevention of scarcity was a step in the right direction.

DECENTRALISED LOCAL AREA PLANNING

Watershed planning must imply decentralisation of the planning process. 'Watershed area planning' would initially call for considerable labour. Both increased employment and higher incomes could result from carefully devised 'watershed area planning' in a village (or cluster of villages). The devolution of power and authority to village *panchayats* would also enthuse the local populace to generate more resources, leading to significantly higher levels of real investment, in turn leading to increased incomes, increased demand and a resultant increased economic activity in backward areas.

Two reviews of the DPAP and the Desert Development Programme (DDP), in 1980 and 1993- 94, have highlighted the lack of political will and the absence of adequate delivery systems as causes for their failure. People's participation, *a sine qua non* of various government and nongovernment development programmes, at the operational level, remains an ill-defined and vague concept. Many programmes whose success depends on the local community's involvement, therefore, begin to flounder after some time.

NGOs all over the country have undertaken several micro-level initiatives for drought proofing through watershed management. Drought proofing has become synonymous with these, without any serious attempt to address the issue of their wider application.

These localised success stories are heavily dependent on NGO initiatives in institutional, financial and technological aspects and are, for many reasons, not immediately amenable to location specific scaling-up in different states of India. It has been difficult to bridge this distance between scale and quality. Policies tend to remain overarching and broad, which, because of their very nature and the attempt to be flexible to suit country-wide variations, end up as vague efforts, and remain ineffective at the level of implementation.

The issue of scaling up NGO efforts is ignored, left to the preambles and directive principles of most drought proofing plans. NGOs too have not given serious thought to this issue. One reason for non-reproducibility of watershed plans is the lack of institutionalisation of people's control and participation.

The 73rd and 74th constitutional amendments provide the opportunity to empower elected local bodies and gram *sabhas* with the requisite finances, functions and administration to become the institutions underpinning drought proofing. We will therefore suggest a state level strategy to empower statutory local bodies with the capabilities for decentralized natural resource management, linking democratization and decentralized governance with sustainability.

In our view drought proofing is a concept, which is not amenable to a simple 'pristine environment' approach. It cannot be achieved simply through water, soil and land conservation. Water scarcity in an iniquitous society has a pronounced social dimension. Landless people and marginal farmers are more vulnerable.

Hence drought-proofing also needs to take account of the factors determining overall water availability and use, within a multidisciplinary approach that combines issues of macroeconomics, agro-ecology, and socio-economic power.

Socio-economic inequity, macroeconomic policy and other issues in political economy, such as condition of the peasantry, balance of power between agriculture and industry, etc. become fundamental in understanding and conceptualising a water policy for drought-proofing, because they in turn affect the vulnerability of the people and the environment, and the possibility or feasibility of institutionalising long-term sustainability and drought-proofing.

Sometimes it is argued that traditional practice is the best and only way to achieve large-scale drought proofing. However, modern scientific and technological advances must be enmeshed with traditional practices through the demystification of modern science and appreciation of indigenous systems, in all their advantages and limitations. Approaches that are either expertbased, or solely emphasize indigenous systems result in the denial of access to modern advances in science and technology and fail to empower people.

Drought proofing offers a meeting point for science and society. For a proper drought-proofing policy, we need to appreciate more fully the holistic approach of traditional practices for a protective irrigation and food production system, which has three components:

- Surface and groundwater management (extraction and recharge) through diversion channels, surface-drainage watershed systems, wells, etc.
- Land reclamation;
- Land use and agricultural practices.

Their simplicity is deceptive - products of generations of experience, they require great expertise in maintenance. The crude structures often form interconnected chains that appropriate every bit of run-off water flowing through vast landscapes for irrigation.

SCALE OF INTERVENTION

The selection of a state as the locale for policy formulation within a federal polity such as India marks a move away from perfectly designed but very small scale and micro interventions, on the one hand, and the necessarily broad and sweeping approach of a national water policy, on the other.

The state is an intermediate unit between the local and the national, which has its own relative autonomy in fiscal and policy matters. More importantly, most functions related to social and economic development rest with the state governments.

While the state offers a tractable unit in functional, administrative and financial terms, it has sub-regional and location-specific variations in agro-ecology, physiology and the biophysical system. Therefore, in order to formulate a significant and relevant state level policy, a smaller biophysical terrain offers an essential input and testing ground for the feasibility and relevance of the policy.

This is made easier with the availability of Geographical Information Systems based computational and processing tools and remote sensing data that significantly increase the ability to handle large database. In other words, a sub-regional planning exercise and the regional policy formulation are iterative rather than distinct processes.

DEVELOPMENT OF UNDERDEVELOPMENT'

The tribals in India are spatially clustered and concentrated in areas, which have suffered from isolation historically and where the environmental setting is by and large unsuited to sedentary agriculture. Indian development has been unsuccessful in integrating tribal areas and people into mainstream development.

The symbiotic relation between the people and their habitat has been ruptured, and pre-existing foundations of tribal livelihoods have been undermined without mitigation or sustainable alternatives. This has led to highly unequal and uneven economic development across regions. Tribal development is the biggest challenge to development planning today. In order to address this challenge, ecologically informed planning is essential.

NEED FOR MULTI-DISCIPLINARY PLANNING

This implies that a uni-dimensional and narrow compartmentalized approach is doomed to fail. The authors of the Atlas are scathing in their critique of the 'anemic tradition of interdisciplinary concerns', and the unique inability of social scientists to cross the *lakshmana rekha* drawn by the colonial tradition is sharply manifested. In fact almost all writings on tribal development, sustainability, natural resources and drought proofing underline the need for multidisciplinary approaches, which are lacking. For instance, in our exercise geology helps understand the sub-surface rock formations and therefore the way aquifers behave and the potential they offer for recharge.

Soil morphology would help arrive at the percolation capacity of the soil and hence determine the water harvesting technology. Depth and extent of moisture retention by the soil can help work out the cropping pattern and irrigation requirement in normal and scarce years, so that biomass potential and output growth can be optimised.

Water treatment, water recycling and low water using process technologies for industry need technologists and engineers. In other words, growth with equity and sustainability requires huge multi-disciplinary efforts.

DROUGHT PRONENESS AND DROUGHT VULNERABILITY

Three concepts that are used distinctively in this work drought, drought vulnerability and drought proneness must be understood clearly. Definitions of drought are enormously complicated, not least because of the huge variety of climatic, geographical, socio-economic and cultural contexts within which a drought occurs.

According to the National Drought Monitoring Centre (NDMC), USA, in most areas that experience drought 'drought is a normal, recurrent feature of climate', that is to say, it is not a rare and random event. They also say that 'drought is a temporary aberration and differs from aridity since the latter is restricted to low rainfall regions and is a permanent feature of climate'.

What constitutes a temporary aberration is vastly different, for example, in areas with a seasonal rainfall pattern (central and north-western India), as compared to areas that have extended periods without rainfall (parts of Rajasthan and Gujarat). The NDMC distinguishes between different definitions of drought: conceptual, operational and disciplinary. The conceptual definition that they offer is 'Drought is a protracted period of deficient precipitation resulting in extensive damage to crops, resulting in loss of yield'.

The disciplinary ones incorporate impacts of deficient rainfall: meteorological, hydrological, agricultural and socio-economic (or ecological). Put simply, they refer, respectively, to rainfall, water levels (in dams, soils, etc.)effects on crop harvest, and effect on communities and the eco-system. The point is that one stage need not necessarily lead to the next stage: a host of factors come into play and intervene. Similarly, the impact of drought can vary according to which 'sector' is looked at.

The impact of drought may diminish rapidly in the agricultural sector because of its reliance on soil water, but linger for months, or even years, in other sectors dependent on stored surface or subsurface supplies.

Ground water users, often the last to be affected by drought during its onset, may be the last to experience a return to normal water levels. In India policy approaches to drought and drought proneness rely on three aspects: rainfall, soil moisture and irrigation. Soil moisture and water balance lie at the heart of how we understand the related concept of dryness and drylands.

Soil moisture depends on several factors: rainfall, temperature, moisture retentivity and porosity of soil, run off of rainwater, vegetation, etc. Drought has been classified into several categories viz., meteorological drought, agricultural drought and hydrological drought. There are two approaches to identifying drought in India, based on the methodology given by the Irrigation Commission and the National Commission on Agriculture.

The Irrigation Commission derives its definition from the Indian Meteorological Department's (IMD) definition of 'drought' as a situation occurring in an area in the year when the rainfall is less then 75 per cent of the normal. 'Moderate drought' is defined as a situation when rainfall deficiency is between 25 to 50 per cent and 'severe drought' when rainfall deficiency is above 50 per cent.

The areas where drought has occurred in at least 20 per cent of the years examined, are classified as 'drought prone areas' and those where drought has occurred in more than 50 per cent of the years examined are considered as 'chronic drought areas'.

This definition of drought does not take into account the effect of irrigation in effective mitigation of the drought on agricultural production. The Irrigation Commission considered both rainfall and irrigation as factors in identifying a drought. It recommended that areas with more than 30 per cent coverage of irrigation should be excluded from being listed as drought prone areas.

The declaration of drought or as a 'scarcity affected area' for providing relief to the population is done by the Revenue Department of the state government, on the basis of estimation of the prospective harvest. The assessment is based on the *'annawari system'* (crop being assessed on the basis of '*annas*').

If the prospective harvest is estimated to be up to 12 *annas* (75 per cent) the crop is normal; and below it up to 6 *annas* (37 per cent), the crop is below normal. A scarcity situation is declared when the crop prospects are less than 6 *annas*. The National Commission on Agriculture, defines drought as 'an occasion when the rainfall in a week is half of the normal or less, when the normal weekly rainfall is 50mm or more.

Agricultural drought is a period of 4 such consecutive weeks in the period from middle of May to middle of October or 6 such consecutive weeks during the rest of the year'. A *drought prone area* is one that has a rainfall history that indicates a high probability of experiencing meteorological drought, which may result in the drying up of groundwater, surface water and soil moisture. If drought proneness describes areas that are likely to face soil moisture deficiency, irrigation must be taken into consideration as it offers important mitigation to shortcomings in precipitation.

Drought prone areas must be those whose water balance is subject to frequent stress that is not mitigated by irrigation. Since availability of irrigation is a crucial determinant of soil moisture, the Drought Prone Area Programme (DPAP) has identified drought prone blocks on the basis of temporal fluctuations around normal rainfall with an irrigation cutoff of 30 per cent of gross sown area.

Such concepts and definitions provide a very useful framework in analysing droughts and in building up a stock of knowledge. However, when employed on their own in specific situations, they are of very limited use in terms of assessing impact on the people, and requirements for effective intervention.

This is so because most of these definitions, and the institutions involved in drought assessment and mitigation, are divorced from the way that populations in drought prone regions experience, conceptualise and respond to droughts. Furthermore, there tends to be a wide gulf between policy makers employing 'objective' methodologies and the reality on the ground.

This difference goes to the heart of the issues involved in drought preparedness and mitigation work that has been done in the country, which is mostly prescriptive in nature, and does little in terms of creating long term assets for the community.

In our view several factors go into the making of a drought, including ecology, production conditions, socioeconomic conditions, etc. Rainfall deficiency (quantum, distribution and reliability) need not necessarily result in distress and shortages. The effect of rainfall deficiency depends on the policy, technology and land-ownership regime in the area and is an outcome of a complex interaction between socio-economic, agro-ecological and policy issues. The ability of the people and the land to withstand rainfall, hydrological and even agricultural deficits depend to a large extent on their status preceding the event or series of events of shortfalls of rain. This status is in turn determined by a large number of factors.

Our points of departure with respect to the above definitions are two fold. We introduce an agrometeorological criterion of drought-proneness to place it on a sounder theoretical footing, by also considering the timeliness and subsequent distribution of rainfall over the season in relation to the cropping pattern of the area.

In addition to the meteorological and soil moisture related predisposition to drought (*drought proneness*) we add the concept of *drought vulnerability*: namely, the ability (or inability) of the land and people to withstand drought or soil moisture distress and experience lower crop failure, out-migration, land alienation, livestock distress, water shortage, hunger and starvation, poor health, etc. . This, in turn, would identify the priority areas for intervention for drought proofing.

Drought Proneness

It is well known that the timeliness, distribution and reliability of rainfall are as important as, if not more so, the total quantum of rainfall. We incorporate these into the definition of drought proneness by taking into account long-term rainfall data on:

- The probable time by which summer monsoon rains in a region would build up enough soil moisture to commence sowing, namely commencement of sowing rain (CSR),
- The inter-spell duration of specified periods during the months of June, July and August to assess and identify the probability of occurrence of next favourable rain spell for the commencement of the re-sowing when the first CSR is missed.

A drought prone area is defined as one, which has a high statistical probability of a consecutive number of intervals where the value of the weekly rainfall throughout kharif lies below a given threshold or truncation level, where the CSR is late and the inter-spell gap is high. This is an important step, which goes beyond the typically mean or normal rainfall analysis of DPAP and IMD.

Drought Vulnerability

Drought vulnerability expresses itself as shortfalls in food, fodder, fuel, water and livelihood. We wanted to identify those most drought prone areas that are highly vulnerable, in the sense, that if rainfall displays any shortcomings with regard to timeliness or quantum, there are one or several reasons why particular areas may suffer greater distress than others.

These may be as seemingly disparate as poverty, landlessness, inadequate infrastructure, high run off rates of groundwater, etc. Rainwater is retained at different rates on the plains and on the hills as soil moisture or groundwater. Hence, the condition of the location where the rain falls is an important consideration in determining interventions.

Rain shadow areas with investment in sound water management strategies and extensive irrigation may experience less crop failure or out-migration than higher rainfall sugarcane growing areas. Grain banks and a good network of the public distribution system may prevent hunger and starvation. High forest and vegetative cover will reduce fodder shortages and livestock loss. Timely commencement of drought relief works and food for work programmes can halt or reduce distress out-migration on account of drought.

Similarly, if groundwater is recharged and aquifers are saved for drinking and domestic use, drinking water crises such as have been witnessed in many areas in the past decade during drought may well become a feature of the past. Land and income transfers to the destitute, the poor and the assetless may go a long way in increasing their ability to withstand the harmful effects of drought. We also wanted to see whether different causes of drought vulnerability resulted in very different types of distress ?crop failure, migration, land alienation, livestock losses, hunger and starvation, poor health, etc.

We argue that drought vulnerability differs for different regions on account of aspects other than meteorological and irrigation performance. In a related exercise, we show that drought is experienced and expressed in different ways, depending upon the determinants of vulnerability.

In some contexts, migration is the predominant expression of distress; in others, it is crop failure. Do different causes of drought vulnerability then necessitate markedly different types of interventions? This is another significant point of departure in this work and leads directly to distinct recommendations, depending upon the expression of drought vulnerability.

ECOLOGICAL TYPOLOGIES

Another important point of departure of our work relates to the categorization of blocks into ecological typologies in order to determine a common matrix of interventions within each typology. Hitherto, exercises in regionalisation have focused essentially on agro-climatic zoning, with landform and terrain as the most important factor that defines or holds together a zone comprising contiguous areas. In a significant departure from this approach, *we have classified and clubbed geographically dispersed areas into common typologies*. This required intensive and meticulous analysis of ecological features (rain, soil, slope, cover, geology), and led to the generation of a common matrix of drought proofing interventions within each typology.

This is unprecedented in its scale, detail and range. It is important for another reason: it demonstrates the possibility and desirability for such an exercise in other states with similarities, like Jharkhand, Orissa, etc in Central and Eastern Peninsular India.

An approach such as the one delineated above generates an extremely rich database for planning that addresses sub-regional specificities and classifies blocks on the basis of drought proneness, drought vulnerability, ecological typologies and drought distress. In each case the approach and treatment marks a step forward from what exists.

Field Survey

It was essential to translate the commonsensical concept of drought vulnerability to meaningful indicators that could be subject to analytical and statistical rigour. We had to observe and understand, in a first hand and direct way, the dynamic and dialectical relationship between ecology, production conditions and state intervention in diverse settings.

In order to unravel the complex interaction between all our variables and identify the most significant indicators, and capture ground realities, we undertook field surveys in three different blocks falling under different typologies in our classification, as well as in different agro-climatic regions of India as identified by the Planning Commission. These were all highly prone and vulnerable to drought.

This gave as an opportunity to identify three features not captured by the secondary data: chronic hunger, indebtedness, land alienation and dispossession reflecting socio-economic deprivation; low infrastructure development, on-farm investment, productivity and wages reflecting inadequate development status; reduction in non-timber forest produce (NTFP) and soil erosion, reflecting environmental degradation.

The field survey also looked at the issues that drought proofing must address in two different broader typologies, namely the plains and the hills. The Scheduled Caste landless labourers of the plains required employment guarantee directly, quite apart from whatever irrigation that may be provided to support *rabi*. In the hills, infrastructure development, protection of *kharif* and increase in agricultural productivity was vital to address the phenomenon of cultivators in poverty. A universal and effective public distribution will be required in both settings. After Independence, the water policy should have been guided more by concerns of development, or at least protection, rather than profit. After all, Chhattisgarh met all the preconditions for the launch of a massive public work's programme for drought proofing.

It had rich potential from tested and reliable traditional systems, which could be most effectively upgraded with the help of modern science and technology. It had soils and crops that had a history of successful irrigation; it had an undulating terrain that offered a rich array of natural sites for small scale impoundment, often with natural embankments and spillways, with a small but vital requirement for constructing masonry sluices and an appropriate distribution system.

Rainfall was by and large plentiful except in a few rain shadow areas, although it was erratic in commencement, distribution and withdrawal. Above all, there was immense need for such a programme, because of the extreme vulnerability of the low-productivity rain-fed agriculture, on the one hand, and the existence of upland tribal farmers dependent on agriculture, on the other.

However, this potential was not realized. However, it must be understood that this continuing neglect of Chhattisgarh's need for protective irrigation in the postcolonial period was for reasons very different from those in the colonial period.

During the Nehruvian period development, rather than profit, guided the investment policy of the State, and there was consensus on the primacy of the State in ushering development. However, due to several interrelated reasons, areas like Chhattisgarh served as the 'periphery' or 'hinterland' and suffered neglect, becoming victims of somebody else's development.

These reasons ranged from political insignificance despite their natural wealth to obsession with infrastructure for large storage and canals that were inappropriate and expensive for these ecologically diverse and fragile drylands; continuing regional disparities and concentration of growth; unwillingness to discipline the rich and mobilise resources for investment; inability to create the decentralized institutional basis for appropriate water resources management, etc.

During the colonial period, the Maikal Hills area in the north was penetrated and exploited far more than the Bastar plateau in the south. After Independence, the money from the Tribal Sub- Plans was pumped into construction of roads in the central part of the Bastar plateau and its isolation gave way to exploitation.

From the Nehru years right up to the nineteen eighties, modernization became synonymous with development, and also became associated with large structures and storages. There was little search for creative alternatives even where this technology was not feasible. At the end of the nineteen eighties, forty years of experience with large structures combined with a rich body of dryland-specific alternatives, once again created the environment for pursuing decentralised and location-specific irrigation.

In 1993, the Seventy Third Amendment provided the missing link — the institutional basis for such a strategy. However, once again this was not to be. By the early nineteen nineties, we had gone round full circle and today, development has once again given way to profit.

The difference is that profitable investment by the colonial state has given way to profitable private investment as the criterion for investment decisions in the irrigation sector. Where this is unlikely, it is assumed that 'self help' will take care. It is to the issues of this debate of the nineteen nineties that we now turn attention.

WATER RESOURCES MANAGEMENT

As we saw in the preceding section, water management has been a hotly contested terrain for a long time. The macroeconomic issues in water resources development that lie at the heart of the contestation are a matter of central concern to this study.

Contributors to the debate have different approaches that stand apart in their theoretical foundations and, hence, in their technical, institutional and financial prescriptions. In the following paragraphs we review the terms of debate and the policy prescriptions flowing from each perspective. This debate has had contributions from broadly four quarters:

- International consultants like Price Waterhouse Coopers and multi-lateral funding agencies like the World Bank and the Asian Development Bank, all of whom represent the anti-state *marketist-privatization* approach;
- Anti-market, anti-state and anti-privatization NGOs and issue-based social movements that represent the *'communitarian'* perspective, wherein the role of state is restricted to financing;
- The *statist* post-Nehruvian advocates of development, who stand for an interventionist, but decentralized and accountable state.
- There is a fourth point of view that is muted and hesitant

and comprises *federalists* who have a very strong developmental perspective, but are pessimistic about the prospects of resource mobilization by the state.

Therefore, they swing in a see-saw between multilateral financial agencies, on the one hand, and NGOs and private players, on the other. For this reason, they are unable to articulate a coherent position with confidence and continue to be sidelined. This is the group to whom we address this study.

AGREEMENT ON SYMPTOMS

There is little disagreement on a few important symptoms of the shortcomings and what has gone wrong in the sector. The first set of issues relates to the *installation, utilization* and *health* or upkeep of water infrastructure. These are:

- Unevenness in the installation and spread of irrigation capacity across the country, as well as within regions;
- Failure to harness even a fifth of the ultimate irrigation potential in states like Chhattisgarh;
- Under-utilization of existing irrigation potential in almost all parts of the country;
- Abysmally poor performance in repair and maintenance due to lack of adequate resources;
- Reduction in the life of structures due to deforestation and the degradation of catchments;
- Shift to water intensive crops in command areas of irrigation projects and overexpolitation of groundwater resources
- Loss of fertile tracts due to waterlogging, salinity, etc.

The second set of issues relates to the fundamental role of water resources development in *agricultural growth* and *food security*. Land use intensity and productivity is positively correlated to the level of irrigation. There is also a high degree of complementarity and connectedness between irrigation and the use of other inputs. Both these associations imply that agricultural production, and consequently, food security depend to a large extent on the availability and extent of irrigation.

The relationship between water and food security is most manifest in the areas of low irrigation: in the peninsular rain-fed dry lands that lack snow-fed rivers and have high run off from steep slopes. Here, soil, water and vegetation are bound together in a dynamic relationship with each other giving rise to the forward and backward linkages between poverty and eco-degradation and are vital determinants of human survival. In such environments traditional and innovative interventions are far more appropriate for irrigation and *in situ* measures. This is so for two reasons: the extreme fragility and location-specificity of the dry lands; and inappropriateness of large storage and groundwater-dependent irrigation structures.

The third set of issues pertains to the failure to institutionalize participatory water resources management and decentralized development, accompanied by neglect of research and development into sustainable dry land technology, for areas where, both, the large reservoir-based single point storage model and the groundwater-dependent green revolution strategy are not techno-economically feasible.

Apart from strictly technical considerations of viability, the widespread displacement and dislocation on account of large dams, largely of tribal farmers, is also a matter of grave concern.

DISAGREEMENT ON SOLUTIONS

However, there are fundamental disagreements regarding of the underlying causes of the existing situation as well as solutions for these. The communitarian and the marketist perspectives begin with a wholesale rejection of the Nehruvian model.

They argue that the Nehruvian interventions in the water sector were problematic on all fronts: the choice of technology in favor of 'modern' and big structures; the huge misery it caused to millions of displaced people and communities, largely in the tribal areas; the primacy it accorded to the state in development planning, financing, delivery and regulation; and the perpetuation of a centralized administrative and institutional machinery with built-in incentives for corrupt and non-accountable governance.

The statists argue that while much was wrong with the *dirigisme*, the central core was based on a national consensus on development through an interventionist state. One of the most fundamental shortcomings of the regime was its inability to carry out land reforms and institute an accountable and democratic state.

This does not in any way undermine the importance of a developmental state; rather, it underlines the importance of democratization of the state, both, through property reform (through land redistribution and progressive taxation) and institutional reform (as envisaged under the *panchayati raj* legislation).

A related issue is the relationship between *public and private investment*. While many argue that public investment has a 'crowding in' effect on private investment and is, in fact, a prerequisite for private investment, neo-classical and neo-liberal economists opine that this

Keynesian insight is wrong and public investment has a 'squeezing out' effect on private investment. This is rooted in the pre-Keynesian notion of a fixed pool of savings. Therefore, neo-classical and neo-liberal economists consider public investment an undesirable policy intervention. The other vexing question, which invokes great passion, is how to achieve *sustainability*.

The relationship between growth, sustainability and equity unfolds, often in markedly different ways, in specific contexts. Sustainability in the use of water would result in a reduction in drought vulnerability. However, the question arises as to what macroeconomic policy environment would be conducive to sustainability, and which one would erode the possibility or feasibility of sustainable development. Moving beyond the reductionist and static view that growth is inimical to sustainability, we are of the view that growth may not be detrimental to sustainability.

The determining factor here is the choice of technology rather than growth *per se*. Often, sustainable development itself is a prerequisite for growth. Furthermore, and this is a thesis that underpins this work right through: *under many circumstances, the absence of growth (through appropriate technology and public works) can itself become a cause of environmental degradation*.

Similarly, inequality itself may be a cause of unsustainable growth. Thus, the reasons behind unsustainable development may be the same as the ones that have stalked these economies and made economic growth difficult, namely, inadequate public investment and absence of adequate re-distributive measures.

This is more so in a poor Third World country like ours, where plunder of natural resources has been the source of accumulation from areas like Chhattisgarh. This would imply that, contrary to the argument of ecological economists like Daly, growth (with equity) might in fact be a prerequisite rather than a detriment inimical to environmental sustainability. One solution is to pursue a strategy of water demand management through market based incentives and disincentives.

The underlying assumption is an elastic demand curve for water, or infinite substitutability between all factors of production, including natural resources. In practice this implies is high *pro-rata* pricing so that prices are high enough to push farmers out of irrigated agriculture. The price at which the disincentives become operational depends on the status of the farmers.

The alternative to this market approach is the argument that all rights should be vested in the community, which has traditional methods of ensuring sustainable water use, which, in turn, can be revived with the help of NGOs.

The third strategy for ensuring sustainable water use involves public investment in decentralized natural resource development through institutions of local self-government; subsidized availability of water saving devices and large-scale support to dry land crops. The policy question then shifts from 'getting the prices right' to what combinations of price and procurement support, input subsidy, credit and technological support can promote an appropriate crop mix. Yet another related set of issues concerns the 'ills' of the green revolution strategy.

Several well-known and now accepted criticisms of the strategy are as follows: it was concentrated in the well-endowed regions and restricted to the better off farmers; the water logging and salinity it caused in many parts of the country where the inherent irrigability of the soil conflicted with the application of water under this strategy; the over extraction of groundwater through energized tubewells in many parts of the country; chemicalization of the soil that destroyed the inherent fertility of the land, etc.

They were undesirable because they resulted in greater inequality and unsustainability. But these are not the concerns of the present critique of the green revolution strategy.

The current criticism is on the *universalized subsidy* for chemical and monetized inputs like fertilizers and pesticides, as well as the non-recovery of water, power tariff, and operation and maintenance costs from users.

The new thinking is that subsidies should be seen as a means of income transfers to the poor, and hence, should be *well targeted*. The counter-argument to this is that the *output responsiveness to irrigation and fertilizer use* is very high, and that application of fertilizers is highly price responsive. Furthermore, the application of fertilizers is also very sensitive and dependent on irrigation.

For this reason universal input subsidy to farmers is desirable for agricultural growth, food security of farmers as well as for national self-sufficiency in food production. In addition, targeting has other problems: leakage through corruption; exclusion of the needy and wrong identification of the poor; precariousness of the 'above poverty line' status of the near poor. It takes very little for many of the non-BPL families to fall into poverty, a phenomenon described as knife-edgedness.

Finally, while poverty figures are under debate, there is little doubt that the per centage of people not meeting their minimum calorific requirement is a shockingly high 70 per cent. The other issue under debate is the reason behind the under-utilization of irrigation potential and low private investment in agriculture. One issue is that the estimates of the installed capacity are themselves suspect, based as they are on all kinds of unrealistic assumptions and norms. There are other critical issues as well.

This discussion has taken on the form of a dispute between *public investment versus good governance* to solve the problem of capacity utilization. Macroeconomic thinking in the Keynesian and Kaleckian tradition has always seen public investment as the prime mover for growth and capacity utilization.

In the Indian context this argument is extended to the creation of a wider economic base of growth through the dispersal and deepening of purchasing power in the hands of the rural and urban poor Agricultural growth is constrained by two factors: public investment in rural works and infrastructure; and inequality and landlessness due to the absence of redistributive measures like land reforms.

Both these features result in the lack of private investment in agriculture and underutilization of irrigation potential in many parts of the country. *This is one of the most contentious issues today, marking the point of departure in the thinking on water policy in this decade.* The international financial institutions that are very influential in policy making today argue that the cause of lack of private investment in agriculture and underutilization of the irrigation potential is not inadequate public investment, but too much public investment, which has squeezed out private investment.

Furthermore, administered prices, including subsidy on savings and cheap credit, have resulted in far reaching market distortions, which again discourage private investment. Bad governance, inefficient delivery systems and unresponsive management all of which can be clubbed under the broad rubric of 'governance' have resulted in underutilization of capacity.

Finally, the highly subsidized tariff structure where even operational and maintenance costs are not recovered has made the entire irrigation sector a grotesque financial black hole. They therefore argue that there are two options for the irrigation sector. One is the present structure with the state playing the central role in financing, provisioning and delivery of water at subsidized rates, with the flip side of inefficiency, delays and unreliability.

The other option is more timely, reliable and efficient irrigation through a combination of the private corporate sector and water users' associations, with the flip side of higher costs and tariffs. In this choice between inefficiency and price, the argument concludes, the farmers prefer higher prices with greater efficiency and reliability.

However, it can also be argued that:

- The state is not inherently or necessarily inefficient and can become more accountable and decentralized,
- That higher prices and private sector participation do not ensure efficiency and timeliness.

There is technological and institutional dualism in the prescriptions put forward by the marketist approach, which explains the somewhat jarring similarity one sometimes finds between the communitarian and the marketist perspective. Large structures for storage of water will be undertaken by the private sector and water resources controlled by them through mega projects like interlinking of rivers, while a few small ventures left to the community's self help with a little seed money from the state.

Both parts of the duality are anti-state, which in any case forms the underlying bondage between the market and the communitarian approaches. However, it is evident that the international financial institutions are promoting two completely contradictory strategies: the 'communitarian' measures with accompanying homilies about traditional knowledge and the highly centralized and gigantic structures with private sector control. Both are guided by the profit motive, and have little to do with people's empowerment. Does irrigation serve the objective of drought proofing or protection in any way? Or is the primacy to irrigation in itself at best peripheral and perhaps even inimical to protection?

It is true that the estimates of utilization of irrigation potential are not achieved, in part, because irrigation brings in its wake a change in cropping patterns, towards more water intensive crops. This phenomenon has given rise to skepticism about irrigation and greater emphasis on 'water demand management'.

The latter is achieved not just through price measures, but also by an exclusive or primary technological focus on soil moisture conservation (in combination with groundwater management) for drought proofing the dry lands, rather than on direct irrigation.

The argument is that groundwater recharge, soil moisture conservation and biomass optimization are undermined if irrigation remains the exclusive focus in an *irrigation versus conservation* conflict. We see them as concomitants and not alternatives, both as part of an integrated and comprehensive water resources strategy.

We look at this issue not so much in terms of 'irrigation versus soil moisture conservation' but one of *choice of technology in specific locations*. Sometimes, in the hills agriculture is on un-bunded fields on sloping land with loose gravelly soils and productivity is extremely low, often not equaling even a third of the state average, which are in any case lower than the Indian figures.

It is argued that such fields are unsuitable for bunded and ploughed cultivation and may do best under a regime of pastoral farming or agroforestry rather than remaining fallow. While this may be correct on purely technical considerations, however, the ends of food security will not be served by such an approach.

Our field survey brought home the fact that this was unacceptable and impractical in places that did not have a reliable and unfailing public distribution system, as well as an effective and functional employment guarantee scheme. A related but dangerous set of questions has re-emerged from the annals of history in the new marketist context, shrouded in the concern for sustainability.

Once again it is being argued that maximizing the returns on investment in water resources development should become the prime mover of investment decisions, not protection or production. That the strategic aim of national food self sufficiency is a mistaken one; it is argued in a neo-Ricardian fashion that temperate countries have an advantage in cereal production.

That we should focus on maximizing the returns to investment in water by growing spices and other high value tropical crops, or unirrigated millets and coarse cereals. Finally, we are told that water productivity is what needs maximization, not land or labour.

We disagree with each one of these prepositions, especially in the context of Chhattisgarh, a backward tribal region of the country where a high proportion of the dispersed working population is engaged in agriculture, as cultivators and as agricultural labourers. Much of this agriculture is under dry and upland conditions.

Therefore, the water policy is closely linked to the spread of growth and livelihood security to the most vulnerable areas in the state. We will argue that dispersed and location specific irrigation, the engine of growth, will serve the twin aims of protection and development.

Apart from small farm ponds, trenches, dykes and farm bunds, protective irrigation in a state like Chhattisgarh, with upland paddy cultivation, will often require storage at lower levels and its lifting when required. Therefore, the availability of affordable and reliable power supply and infrastructure is an important concomitant of any drought proofing strategy.

We will argue that both state-provisioned irrigation and *panchayat*-led watershed development for soil moisture conservation must necessarily be a part of any comprehensive drought proofing water policy for Chhattisgarh.

An excursion through the specificities of Chhattisgarh and its sub-regional characteristics. It presents the *block-level ecological profile, production conditions and socio-economic* characteristics of the state. The ecological profile includes total *rainfall* and its distribution and reliability; *geology* and groundwater potential; *soil* particle size, depth, drainage, erosion, etc; *slope* and *landform*; vegetation and *forest cover*; watershed and *drainage* characteristics to understand run off rates; etc.

Production conditions include *yield, cropping pattern,* source-wise *irrigation, groundwater potential, utilization* of ultimate and existing *irrigation potential, land use patterns.* etc. *Socio-economic status* looks at *land distribution, poverty, workforce* characteristics, composition by *caste* and *tribe*, etc.

The second section presents the statistical exercises conducted on the secondary data. The exercise on identification of the most drought prone and drought vulnerable blocks of Chhattisgarh follow this. The ecological sub-regionalisation comes next, wherein the blocks have been categorised into different *typologies* amenable to a common matrix of interventions.

The final section presents the *field survey* in three distinct typologies spread over twenty odd villages. The final section is a *policy review* of the water sector in the region, where we look at the historical debates on water resources development for drought proofing agriculture in Chhattisgarh. This discussion is embedded in the overall macroeconomic policy environment within which the specific sectoral policies are set.

It examines the *role of state*; the prospects of *profit* as the guiding principle for investment decisions in water works; the *potential of effective drought proofing* by mitigating natural shortfalls through water management strategies; the importance of *power resources* development for water management in the hills; debates around identification of technically feasible solutions; plough versus broadcast in the uplands; pastoral and agro-forestry versus upland paddy, coarse cereals and millets in the light soils on sloping hills; etc.

The existing policy, financial allocations and performance are examined critically in order to identify lacunae and suggest specific interventions and their financing.

We conclude by suggesting a water policy and implementation strategy for Chhattisgarh. In particular, we work out an employment guarantee scheme for the impoverished Scheduled Caste labourers in the plains and a labour-intensive, watershed based drought-proofing package for the tribal cultivators living in penury in the hills. It is our hope that this work will pose a challenge to the other ideologues identified earlier on.

Namely, those who put forth neo-liberal policies that rest on the erroneous theoretical foundations of neo-Ricardian comparative advantage, which argues against food self sufficiency in favour of high value tropical spices and aromatics; those who propose pre-Keynesian contractionary fiscal policies that argue against state intervention and public investment; and those for neo-Malthusian ecological macro-economics that pits growth against conservation with price as the only saviour.

We hope that the state government will revert to its correct position adopted at the time of state formation and abandon the marketist prescriptions of Price Waterhouse Coopers and others of their ilk. The prescriptions flowing from the market-based approach are doomed to fail and the promise of development of the region will remain a chimera.

Chapter 3

Microbial Biodegradation

Interest in the microbial biodegradation of pollutants has intensified in recent years as humanity strives to find sustainable ways to cleanup contaminated environments.

These bioremediation and biotransformation methods endeavour to harness the astonishing, naturally occurring, ability of microbial xenobiotic metabolism to degrade, transform or accumulate a huge range of compounds including hydrocarbons (e. g. oil), polychlorinated biphenyls (PCBs), polyaromatic hydrocarbons (PAHs), pharmaceutical substances, radionuclides and metals.

Major methodological breakthroughs in recent years have enabled detailed genomic, metagenomic, proteomic, bioinformatic and other high-throughput analyses of environmentally relevant microorganisms providing unprecedented insights into key biodegradative pathways and the ability of organisms to adapt to changing environmental conditions.

The elimination of a wide range of pollutants and wastes from the environment is an absolute requirement to promote a sustainable development of our society with low environmental impact. Biological processes play a major role in the removal of contaminants and they take advantage of the astonishing catabolic versatility of microorganisms to degrade/convert such compounds.

New methodological breakthroughs in sequencing, genomics, proteomics, bioinformatics and imaging are producing vast amounts of information. In the field of Environmental Microbiology, genome-based global studies open a new era providing unprecedented *in silico* views of metabolic and regulatory networks, as well as clues to the evolution of degradation pathways and to the molecular adaptation strategies to changing environmental conditions.

Functional genomic and metagenomic approaches are increasing our understanding of the relative importance of different pathways and regulatory networks to carbon flux in particular environments and for particular compcunds and they will certainly accelerate the development of bioremediation technologies and biotransformation processes.

Aerobic Biodegradation of Pollutants

The burgeoning amount of bacterial genomic data provides unparalleled opportunities for understanding the genetic and molecular bases of the degradation of organic pollutants. Aromatic compounds are among the most recalcitrant of these pollutants and lessons can be learned from the recent genomic studies of *Burkholderia xenovorans* LB400 and *Rhodococcus* sp. strain RHA1, two of the largest bacterial genomes completely sequenced to date.

These studies have helped expand our understanding of bacterial catabolism, non-catabolic physiological adaptation to organic compounds, and the evolution of large bacterial genomes. First, the metabolic pathways from phylogenetically diverse isolates are very similar with respect to overall organization.

Thus, as originally noted in pseudomonads, a large number of "peripheral aromatic" pathways funnel a range of natural and xenobiotic compounds into a restricted number of "central aromatic" pathways. Nevertheless, these pathways are genetically organized in genus-specific fashions, as exemplified by the b-ketoadipate and Paa pathways. Comparative genomic studies further reveal that some pathways are more widespread than initially thought. Thus, the Box and Paa pathways illustrate the prevalence of non-oxygenolytic ring-cleavage strategies in aerobic aromatic degradation processes. Functional genomic studies have been useful in establishing that even organisms harboring high numbers of homologous enzymes seem to contain few examples of true redundancy.

For example, the multiplicity of ring-cleaving dioxygenases in certain rhodococcal isolates may be attributed to the cryptic aromatic catabolism of different terpenoids and steroids. Finally, analyses have indicated that recent genetic flux appears to have played a more significant role in the evolution of some large genomes, such as LB400's, than others.

However, the emerging trend is that the large gene repertoires of potent pollutant degraders such as LB400 and RHA1 have evolved principally through more ancient processes. That this is true in such phylogenetically diverse species is remarkable and further suggests the ancient origin of this catabolic capacity.

Anaerobic Biodegradation of Pollutants

Anaerobic microbial mineralization of recalcitrant organic

pollutants is of great environmental significance and involves intriguing novel biochemical reactions. In particular, hydrocarbons and halogenated compounds have long been doubted to be degradable in the absence of oxygen, but the isolation of hitherto unknown anaerobic hydrocarbon-degrading and reductively dehalogenating bacteria during the last decades provided ultimate proof for these processes in nature. Many novel biochemical reactions were discovered enabling the respective metabolic pathways, but progress in the molecular understanding of these bacteria was rather slow, since genetic systems are not readily applicable for most of them. However, with the increasing application of genomics in the field of environmental microbiology, a new and promising perspective is now at hand to obtain molecular insights into these new metabolic properties.

Several complete genome sequences were determined during the last few years from bacteria capable of anaerobic organic pollutant degradation. The ~4. 7 Mb genome of the facultative denitrifying *Aromatoleum aromaticum* strain EbN1 was the first to be determined for an anaerobic hydrocarbon degrader (using toluene or ethylbenzene as substrates).

The genome sequence revealed about two dozen gene clusters (including several paralogs) coding for a complex catabolic network for anaerobic and aerobic degradation of aromatic compounds. The genome sequence forms the basis for current detailed studies on regulation of pathways and enzyme structures. Further genomes of anaerobic hydrocarbon degrading bacteria were recently completed for the iron-reducing species *Geobacter metallireducens* and the perchlorate-reducing *Dechloromonas aromatica,* but these are not yet evaluated in formal publications.

Complete genomes were also determined for bacteria capable of anaerobic degradation of halogenated hydrocarbons by halorespiration: the ~1. 4 Mb genomes of *Dehalococcoides ethenogenes* strain 195 and *Dehalococcoides* sp. strain CBDB1 and the ~5. 7 Mb genome of *Desulfitobacterium hafniense* strain Y51. Characteristic for all these bacteria is the presence of multiple paralogous genes for reductive dehalogenases, implicating a wider dehalogenating spectrum of the organisms than previously known.

Moreover, genome sequences provided unprecedented insights into the evolution of reductive dehalogenation and differing strategies for niche adaptation.

Bioavailability, Chemotaxis, and Transport of Pollutants

Bioavailability, or the amount of a substance that is

physiochemically accessible to microorganisms is a key factor in the efficient biodegradation of pollutants.

Chemotaxis, or the directed movement of motile organisms towards or away from chemicals in the environment is an important physiological response that may contribute to effective catabolism of molecules in the environment. In addition, mechanisms for the intracellular accumulation of aromatic molecules via various transport mechanisms are also important.

Oil Biodegradation

Petroleum oil contains aromatic compounds that are toxic for most life forms. Episodic and chronic pollution of the environment by oil causes major ecological perturbations. Marine environments are especially vulnerable since oil spills of coastal regions and the open sea are poorly containable and mitigation is difficult. In addition to pollution through human activities, millions of tons of petroleum enter the marine environment every year from natural seepages.

Despite its toxicity, a considerable fraction of petroleum oil entering marine systems is eliminated by the hydrocarbon-degrading activities of microbial communities, in particular by a remarkable recently discovered group of specialists, the so-called hydrocarbonoclastic bacteria (HCB). *Alcanivorax borkumensis* was the first HCB to have its genome sequenced.

Analysis of Waste Biotreatment

Sustainable development requires the promotion of environmental management and a constant search for new technologies to treat vast quantities of wastes generated by increasing anthropogenic activities. Biotreatment, the processing of wastes using living organisms, is an environmentally friendly, relatively simple and cost-effective alternative to physico-chemical clean-up options.

Confined environments, such as bioreactors], have been engineered to overcome the physical, chemical and biological limiting factors of biotreatment processes in highly controlled systems. The great versatility in the design of confined environments allows the treatment of a wide range of wastes under optimized conditions.

To perform a correct assessment, it is necessary to consider various microorganisms having a variety of genomes and expressed transcripts and proteins.

A great number of analyses are often required. Using traditional genomic techniques, such assessments are limited and time-consuming. However, several high-throughput techniques originally developed for medical studies can be applied to assess biotreatment in confined environments.

Metabolic Engineering and Biocatalytic Applications

The study of the fate of persistent organic chemicals in the environment has revealed a large reservoir of enzymatic reactions with a large potential in preparative organic synthesis, which has already been exploited for a number of oxygenases on pilot and even on industrial scale. Novel catalysts can be obtained from metagenomic libraries and DNA sequence based approaches. Our increasing capabilities in adapting the catalysts to specific reactions and process requirements by rational and random mutagenesis broadens the scope for application in the fine chemical industry, but also in the field of biodegradation. In many cases, these catalysts need to be exploited in whole cell bioconversions or in fermentations, calling for system-wide approaches to understanding strain physiology and metabolism and rational approaches to the engineering of whole cells as they are increasingly put forward in the area of systems biotechnology and synthetic biology.

CHEMOTAXIS

Chemotaxis, a kind of taxis, is the phenomenon in which bodily cells, bacteria, and other single-cell or multicellular organisms direct their movements according to certain chemicals in their environment. This is important for bacteria to find food (for example, glucose) by swimming towards the highest concentration of food molecules, or to flee from poisons (for example, phenol).

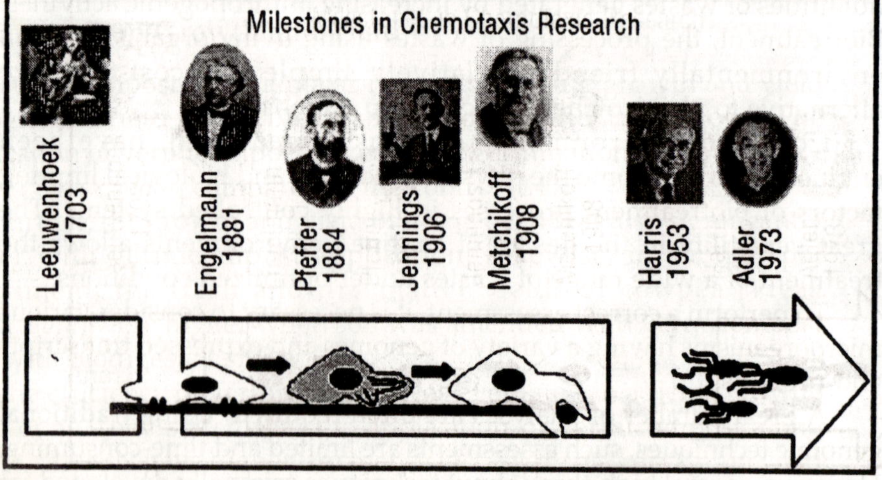

Fig. Milestones in Cheotaxis

In multicellular organisms, chemotaxis is critical to early (e. g. movement of sperm towards the egg during fertilization) and subsequent phases of development (e. g. migration of neurons or lymphocytes) as well as in normal function. In addition, it has been recognized that mechanisms that allow chemotaxis in animals can be subverted during cancer metastasis.

Chemotaxis is called positive if movement is in the direction of a higher concentration of the chemical in question, and negative if the direction is opposite.

Milestones of Chemotaxis Research

Although migration of cells was detected from the early days of the development of microscopy *(Leeuwenhoek)*, erudite description of chemotaxis was first made by *T. W. Engelmann* (1881) and *W. F. Pfeffer* (1884) in bacteria and *H. S. Jennings* (1906) in ciliates. The Nobel Prize Laureate *E. Metchnikoff* also contributed to the study of the field with investigations of the process as an initial step of phagocytosis.

The significance of chemotaxis in biology and clinical pathology was widely accepted in the 1930s. The most fundamental definitions belonging to the phenomenon were also drafted by this time. The most important aspects in quality control of chemotaxis assays were described by *H. Harris* in the 1950s.

In the 1960s and 1970s, the revolution of modern cell biology and biochemistry provided a series of novel techniques which became available to investigate the migratory responder cells and subcellular fractions responsible for chemotactic activity.

The pioneering works of *J. Adler* represented a significant turning point in understanding the whole process of intracellular signal transduction of bacteria.

Phylogeny and Chemotactic Signaling

Chemotaxis is one of the most basic cell physiological responses. Development of receptor systems for the detection of harmful and favorable substances in the environment was most essential to unicellular organisms from the very early stages of phylogeny. Comprehensive analysis of chemotactic activity of the eukaryotic protozoon *Tetrahymena pyriformis* and consensus sequences of appearance of amino acids in the primordial soup suggest that there was a good correlation between the chemotactic character of these relative simple organic molecules and their development on the Earth.

In this way the earliest molecules are suggested to be highly chemoattractant (e. g. Gly, Glu, Pro), while latter ones are thought to be strongly chemorepellent (e. g. Tyr, Trp, Phe) amino acids.

Bacterial Chemotaxis

Some bacteria, such as *E. coli*, have several flagella per cell (4–10 typically). These can rotate in two ways :

- Counter-clockwise rotation aligns the flagella into a single rotating bundle, causing the bacterium to swim in a straight line.
- Clockwise rotation breaks the flagella bundle apart such that each flagellum points in a different direction, causing the bacterium to tumble in place.

The directions of rotation are given for an observer outside the cell looking down the flagella toward the cell.

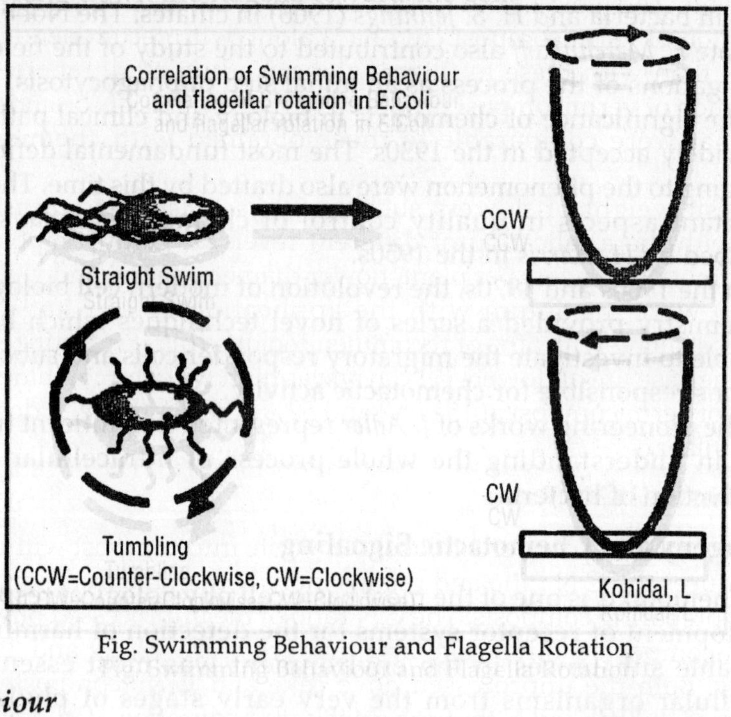

Fig. Swimming Behaviour and Flagella Rotation

Behaviour

The overall movement of a bacterium is the result of alternating tumble and swim phases. If one watches a bacterium swimming in a uniform environment, its movement will look like a random walk with relatively straight swims interrupted by random tumbles that reorient the bacterium.

Bacteria such as *E. coli* are unable to choose the direction in which they swim, and are unable to swim in a straight line for more than a few seconds due to rotational diffusion. In other words, bacteria "forget" the direction in which they are going. Given these limitations, it is remarkable that bacteria can direct their motion to find favorable locations with high concentrations of attractants (usually food) and avoid repellents (usually poisons).

In the presence of a chemical gradient bacteria will chemotax, or direct their overall motion based on the gradient. If the bacterium senses that it is moving in the correct direction (toward attractant/away from repellent), it will keep swimming in a straight line for a longer time before tumbling. If it is moving in the wrong direction, it will tumble sooner and try a new direction at random.

In other words, bacteria like *E. coli* use temporal sensing to decide whether life is getting better or worse. In this way, it finds the location with the highest concentration of attractant (usually the source) quite well. Even under very high concentrations, it can still distinguish very small differences in concentration. Fleeing from a repellent works with the same efficiency.

It seems remarkable that this purposeful random walk is a result of simply choosing between two methods of random movement; namely tumbling and straight swimming. In fact, chemotactic responses such as *forgetting* direction and *choosing* movements resemble the decision-making abilities of higher lifeforms with brains that process sensory data. The helical nature of the individual flagellar filament is critical for this movement to occur. As such, the protein that makes up the flagellar filament, flagellin, is quite similar among all flagellated bacteria. Vertebrates seem to have taken advantage of this fact by possessing an immune receptor (TLR5) designed to recognize this conserved protein.

As in many instances in biology, there are bacteria that do not follow this rule. Many bacteria, such as *Vibrio*, are monoflagellated and have a single flagellum at one pole of the cell. Their method of chemotaxis is different. Others possess a single flagellum that is kept inside the cell wall. These bacteria move by spinning the whole cell, which is shaped like a corkscrew.

Signal Transduction

Chemical gradients are sensed through multiple transmembrane receptors, called methyl accepting chemotaxis proteins (MCPs), which vary in the molecules that they detect. These receptors may bind attractants or repellents directly or indirectly through interaction with proteins of periplasmatic space. The signals from these receptors are transmitted across the plasma membrane into the cytosol, where *Che proteins* are activated. The Che proteins alter the tumbling frequency, and alter the receptors.

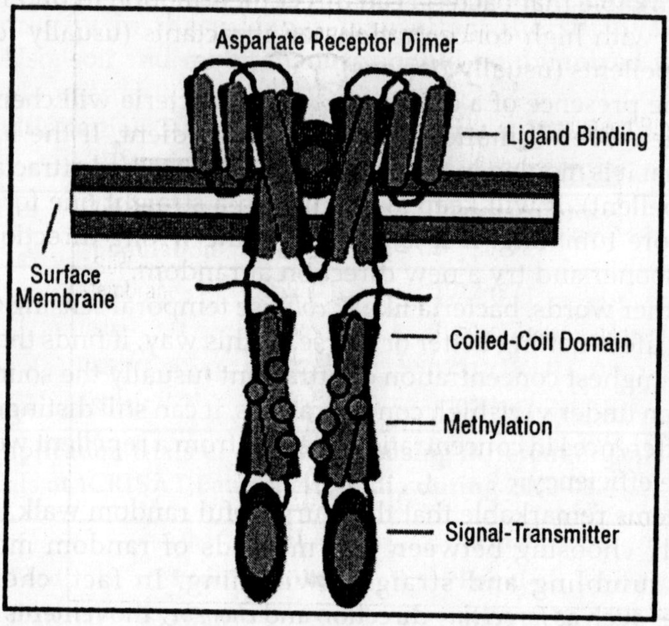

Fig. Signal Transduction

Flagellum Regulation

The proteins CheW and CheA bind to the receptor. The activation of the receptor by an external stimulus causes autophosphorylation in the histidine kinase, CheA, at a single highly conserved histidine residue. CheA in turn transfers phosphoryl groups to conserved aspartate residues in the response regulators CheB and CheY [note: CheA is a histidine kinase and it does not actively transfer the phosphoryl group.

The response regulator CheB takes the phosphoryl group from CheA]. This mechanism of signal transduction is called a two-component system and is a common form of signal transduction in bacteria. CheY induces tumbling by interacting with the flagellar switch protein FliM, inducing a change from counter-clockwise to clockwise rotation of the flagellum. Change in the rotation state of a single flagellum can disrupt the entire flagella bundle and cause a tumble.

Receptor Regulation

CheB, when activated by CheA, acts as a methylesterase, removing methyl groups from glutamate residues on the cytosolic side of the receptor. It works antagonistically with CheR, a methyltransferase, which adds methyl residues to the same glutamate residues.

The more methyl residues are attached to the receptor, the more sensitive the receptor. As the signal from the receptor induces demethylation of the receptor in a feedback loop, the system is continuously adjusted to environmental chemical levels, remaining sensitive for small changes even under extreme chemical concentrations.

This regulation allows the bacterium to 'remember' chemical concentrations from the recent past, a few seconds, and compare them to those it is currently experiencing, thus 'know' whether it is traveling up or down a gradient. However, the methylation system alone cannot account for the wide range of sensitivity that bacteria have to chemical gradients. Additional regulatory mechanisms such as receptor clustering and receptor-receptor interactions also modulate the signalling pathway.

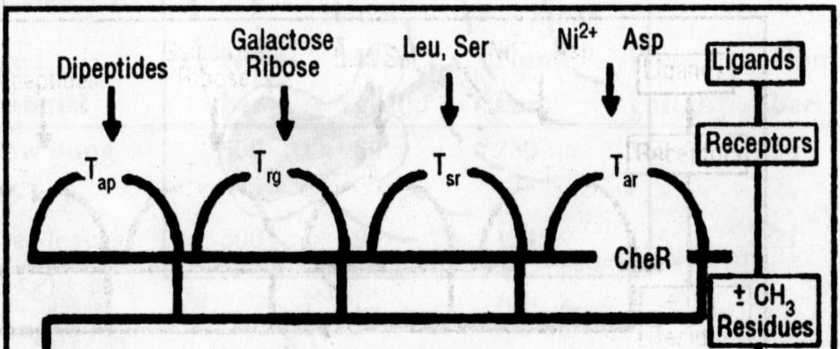

Fig. Receptor Regulation

Eukaryotic Chemotaxis

The mechanism by which eukaryotic cells chemotax is quite different from that in bacteria; however, sensing of chemical gradients is still a crucial step in the process. Due to their size, prokaryotes cannot detect effective concentration gradients, therefore these cells scan and evaluate their environment by a constant swimming (consecutive steps of straight swims and tumbles).

In contrast to prokaryotes, the size of eukaryotic cells allows for the possibility of detecting gradients, which results in a dynamic and polarized distribution of receptors. Induction of these receptors by chemoattractants or chemorepellents results in migration towards or away from the chemotactic substance.

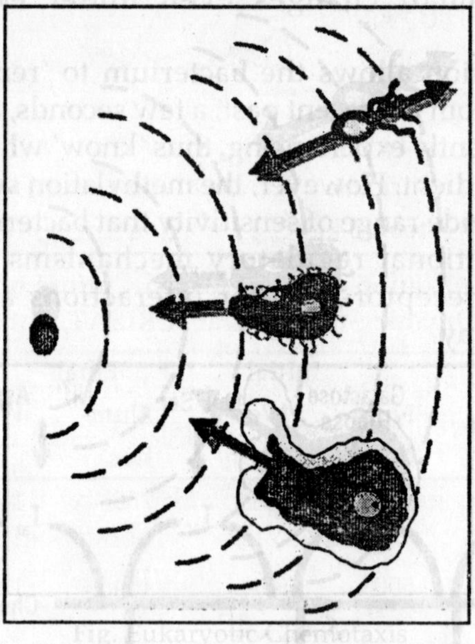

Fig. Eukaryotic Chemotaxis

Levels of receptors, intracellular signalling pathways and the effector mechanisms all represent diverse, eukaryotic type components. In eukaryotic unicellular cells, ameboid movement and cilium or the eukaryotic flagellum are the main effectors (e. g. Amoeba or Tetrahymena). Some eukaryotic cells of higher vertebrate origin, such as immune cells also move to where they need to be.

Besides immune competent cells (granulocyte, monocyte, lymphocyte) a large group of cells - considered previously to be fixed into tissues - are also motile in special physiological (e. g. mast cell, fibroblast, endothelial cells)or pathological conditions (e. g. metastases). Chemotaxis has high significance in the early phases of embryogenesis as development of germ layers is guided by gradients of signal molecules.

Motility

Unlike motility in bacterial chemotaxis, the mechanism by which eukaryotic cells physically move is unclear. There appear to be mechanisms by which an external chemotactic gradient is sensed and turned into an intracellular PIP3 gradient, which results in a gradient and the activation of a signaling pathway, culminating in the polymerisation of actin filaments.

The growing distal end of actin filaments develops connections with the internal surface of the plasma membrane via different sets of peptides and results in the formation of pseudopods. Cilia of eukaryotic cells can also produce chemotaxis; in this case it is mainly a Ca^{2+} dependent induction of the microtubular system of the basal body and the beat of the 9+2 microtubules within cilia.

The orchestered beating of hundreds of cilia is synchronized by a submembranous system built between basal bodies. The details of the signaling pathways are still not totally clear.

CHEMOTAXIS RELATED MIGRATORY RESPONSES

Although chemotaxis is the most frequently studied form of migration there are several other forms of locomotion in the cellular level.

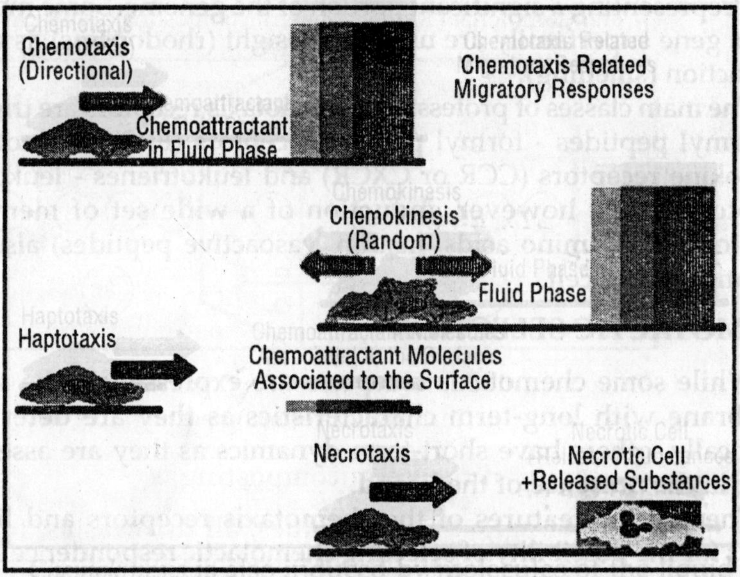

- *Chemokinesis* is also induced by molecules of the liquid phase of the surrounding environment; however, the response elicited is a not vectorial, random taxis. Neither amplitude nor frequency of motion has characteristic, directional components as this behaviour provides more scanning of the environment than migration between two distinct points.

- In *haptotaxis* the gradient of the chemoattractant is expressed or bound on a surface, in contrast to the classical model of chemotaxis, in which the gradient develops in a soluble fluid.
  The most common biologically active haptotactic surface is the extracellular matrix (ECM); the presence of bound ligands is responsible for induction of transendothelial migration and angiogenesis.
- *Necrotaxis* embodies a special type of chemotaxis when the chemoattractant molecules are released from necrotic or apoptotic cells. Depending on the chemical character of released substances, necrotaxis can accumulate or repel cells, which underlines the pathophysiological significance of this phenomenon.

RECEPTORS

For the most part, eukaryotic cells sense the presence of chemotactic stimuli though the use of 7-transmembrane (or serpentine) heterotrimeric G-protein coupled receptors. This class of receptors is huge, representing a significant portion of the genome. Some members of this gene superfamily are used in eyesight (rhodopsins) as well as in olfaction (smelling).

The main classes of professional chemotaxis receptors are triggered by formyl peptides - formyl peptide receptors (FPR), chemokines - chemokine receptors (CCR or CXCR) and leukotrienes - leukotriene receptors (BLT); however, induction of a wide set of membrane receptors (e. g. amino acids, insulin, vasoactive peptides) also elicit migration of the cell.

CHEMOTACTIC SELECTION

While some chemotaxis receptors are expressed in the surface membrane with long-term characteristics as they are determined genetically, others have short-term dynamics as they are assembled *ad hoc* in the presence of the ligand.

The diverse features of the chemotaxis receptors and ligands allows for the possibility of selecting chemotactic responder cells with a simple chemotaxis assay. By chemotactic selection we can determine whether a still uncharacterized molecule acts via the long- or the short-term receptor pathway.

The term *chemotactic selection* is also used to designate a technique which separates eukaryotic or prokaryotic cells according to their chemotactic responsiveness to selector ligands.

CHEMOTACTIC LIGANDS

The number of molecules capable of eliciting chemotactic responses is relatively high, and we can distinguish primary and secondary chemotactic molecules. The main groups of the primary ligands are as follows:

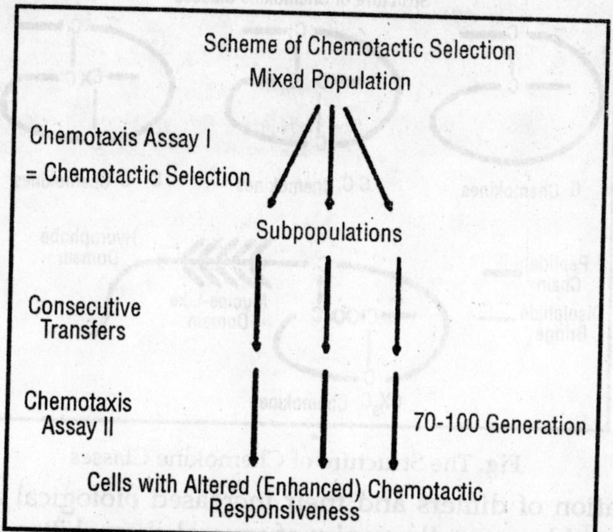

Fig. Groups of the Primary Ligands

- *Formyl peptides* are di-, tri-, tetrapeptides of bacterial origin. They are released from bacteria in vivo or after decomposition of the cell. A typical member of this group is the N-formylmethionyl-leucyl-phenylalanine. The bacterial origin fMLF as a key component of inflammation has characteristic chemoattractant effects in neutrophil granulocytes and monocytes.
- *Complement 3a (C3a) and complement 5a (C5a)* are intermediate products of complement cascade. Their synthesis is joined to the three alternative pathways (classical, lectin dependent and alternative) of complement activation by a convertase enzyme. The main target cells of these derivaties are neutrophil granulocytes and monocytes as well.
- *Chemokines* belong to a special class of cytokines. Their groups represent not only structurally related molecules with a special arrangement of disulfide bridges, but their target cell specificity is also diverse: CC chemokines are acting on monocytes (e. g. RANTES), CXC chemokines are neutrophil granulocyte specific.

Investigations of the three-dimensional structures of chemokines proved that a characteristic composition of beta-sheets and an alpha helix provides expression of sequence s required for interaction with the chemokine receptors.

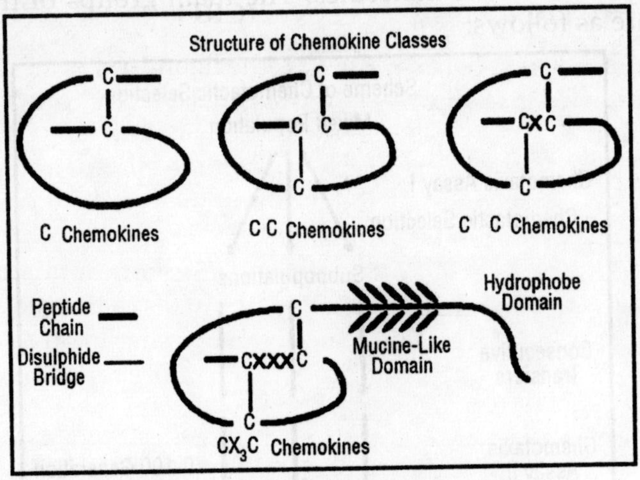

Fig. The Structure of Chemokine Classes

Formation of dimers and their increased biological acitvity was demonstrated by crystallography of several chemokines e. g. IL-8.

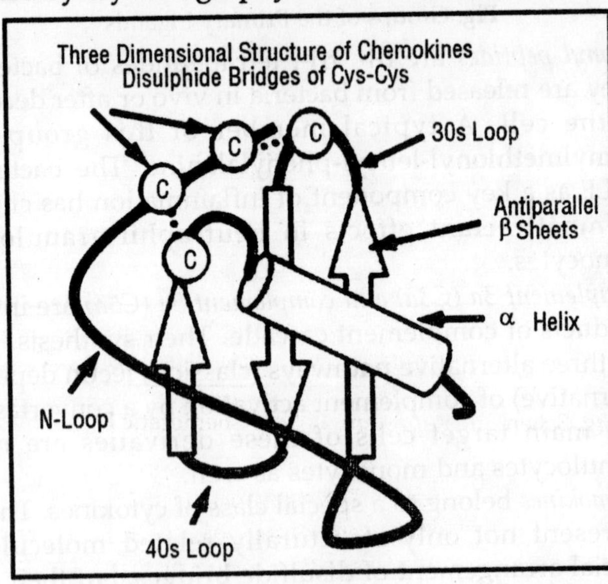

Fig. Three Dimensional Structure of Chemokines

Leukotrienes belong to the group eicosanoids. They are significant ipid mediators of the arachidonic acid cascade converted by 5-

lipoxigenase. Their predominant member is leukotriene B4 (LTB4) which elicits adhesion, chemotaxis and aggregation of leukocytes.

The characteristic chemoattractant effect of LTB4 is induced via G-protein linked seven-transmembrane spanning leukotriene receptors which are highly expressed in inflammation and allergy.

CHEMOTACTIC RANGE FITTING (CRF)

Chemotactic responses elicited by the ligand-receptor interactions are distinguished generally upon the optimal effective concentration(s) of the ligand.

Nevertheless, correlation of the amplitude elicited and ratio of the responder cells compared to the total number are also characteristic features of the chemotactic signaling.

Investigations of ligand families (e. g. amino acids or oligo peptides) proved that there is a fitting of ranges (amplitudes; number of responder cells) and chemotactic activities: chemoattractant moiety is accompanied by wide ranges, while chemorepellent character by narrow ranges.

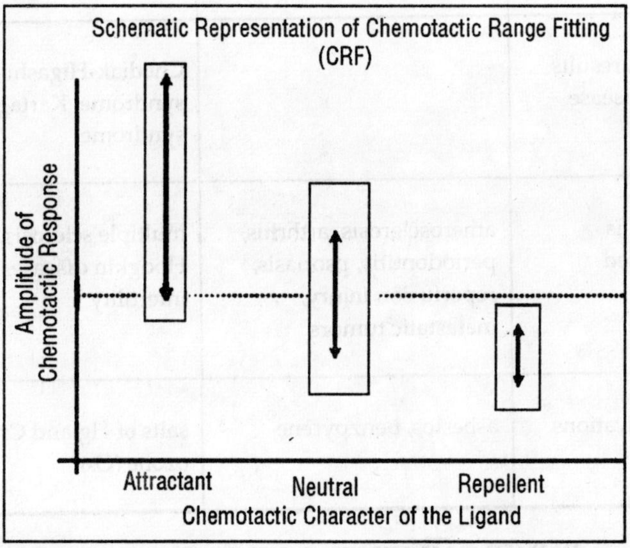

Fig. Schematic Representation of Chemotatic Range Fitting

CLINICAL SIGNIFICANCE

A changed migratory potential of cells has relatively high importance in the development of several clinical symptoms and syndromes.

Altered chemotactic activity of extracellular (e. g. *Escherichia coli*)

or intracellular (e. g. *Listeria monocytogenes*) pathogens itself represents a significant clinical target.

Modification of endogenous chemotactic ability of these microorganisms by pharmaceutical agents can decrease or inhibit the ratio of infections or spreading of infectious diseases.

Apart from infections, there are some other diseases where impaired chemotaxis is the primary etiological factor, as in Chediak-Higashi syndrome where giant intracellular vesicles inhibit normal migration of cells.

Chemotaxis (Chtx.) in diseases		
Type of disease	Chtx. increased	Chtx. decreased
Infections	inflammations	AIDS, Brucellosis
Chtx. results the disease	-	Chediak-Higashi syndrome, Kartagener syndrome
Chtx. is affected	atherosclerosis, arthritis, periodontitis, psoriasis, reperfusion injury, metastatic tumors	multiple sclerosis, Hodgkin disease, male infertility
Intoxications	asbestos, benzpyrene	salts of Hg and Cr, ozone (O_3)

ENVIRONMENTAL MICROBIOLOGY

Environmental microbiology is the study of the composition and physiology of microbial communities in the environment. The environment in this case means the soil, water, air and sediments covering the planet and can also include the animals and plants that inhabit these areas. Environmental microbiology also includes the study of microorganisms that exist in artificial environments such as bioreactors. Microbial life is amazingly diverse and microorganisms literally cover the planet. It is estimated that we know fewer than 1% of the microbial species on Earth. Microorganisms can survive in some of the most extreme environments on the planet and some, for example the Archaea, can survive high temperatures, often above 100°C, as found in geysers, black smokers, and oil wells. Some are found in very cold habitats and others in highly saline, acidic, or alkaline water.

An average gram of soil contains approximately one billion (1,000,000,000) microbes representing probably several thousand species. Microorganisms have special impact on the whole biosphere. They are the backbone of ecosystems of the zones where light cannot approach.

In such zones, chemosynthetic bacteria are present which provide energy and carbon to the other organisms there. Some microbes are decomposers which have ability to recycle the nutrients.

Microbes have a special role in biogeochemical cycles. Microbes, especially bacteria, are of great importance because their symbiotic relationship (either positive or negative) have special effects on the ecosystem. Microorganisms are used for *in-situ* microbial biodegradation or bioremediation of domestic, agricultural and industrial wastes and subsurface pollution in soils, sediments and marine environments.

The ability of each microorganism to degrade toxic waste depends on the nature of each contaminant. Since most sites typically have multiple pollutant types, the most effective approach to microbial biodegradation is to use a mixture of bacterial species and strains, each specific to the biodegradation of one or more types of contaminants.

It is vital to monitor the composition of the indigenous and added bacteria in order to evaluate the activity level and to permit modifications of the nutrients and other conditions for optimizing the bioremediation process.

OIL BIODEGRADATION

Petroleum oil is toxic, and pollution of the environment by oil causes major ecological concern. Oil spills of coastal regions and the open sea are poorly containable and mitigation is difficult; much of the oil can, however, be eliminated by the hydrocarbon-degrading activities of microbial communities, in particular the hydrocarbonoclastic bacteria (HCB).

These organisms can help remedy the ecological damage caused by oil pollution of marine habitats. HCB also have potential biotechnological applications in the areas of bioplastics and biocatalysis.

DEGRADATION OF AROMATIC COMPOUNDS BY ACINETOBACTER

Acinetobacter strains isolated from the environment are capable of the biodegradation of a wide range of aromatic compounds. The predominant route for the final stages of assimilation to central metabolites is through catechol or protocatechuate (3,4-dihydroxybenzoate) and the beta-ketoadipate pathway and the diversity within the genus lies in the channelling of growth substrates, most of which are natural products of plant origin, into this pathway.

ANALYSIS OF WASTE BIOTREATMENT

Biotreatment, the processing of wastes using living organisms, is an environmentally friendly alternative to other options for treating waste material. Bioreactors] have been designed to overcome the various limiting factors of biotreatment processes in highly controlled systems.

This versatility in the design of bioreactors allows the treatment of a wide range of wastes under optimized conditions. It is vital to consider various microorganisms and a great number of analyses are often required.

ENVIRONMENTAL GENOMICS OF CYANOBACTERIA

The application of molecular biology and genomics to environmental microbiology has led to the discovery of a huge complexity in natural communities of microbes. Diversity surveying, community fingerprinting and functional interrogation of natural populations have become common, enabled by a range of molecular and bioinformatics techniques.

Recent studies on the ecology of Cyanobacteria have covered many habitats and have demonstrated that cyanobacterial communities tend to be habitat-specific and that much genetic diversity is concealed among morphologically simple types. Molecular, bioinformatics, physiological and geochemical techniques have combined in the study of natural communities of these bacteria.

CORYNEBACTERIA

Corynebacteria are a diverse group Gram-positive bacteria found in a range of different ecological niches such as soil, vegetables, sewage, skin, and cheese smear. Some, such as *Corynebacterium diphtheriae,* are important pathogens while others, such as *Corynebacterium glutamicum,* are of immense industrial importance. *C. glutamicum* is one of the biotechnologically most important bacterial species with an annual production of more than two million tons of amino acids, mainly L-glutamate and L-lysine.

LEGIONELLA

Legionella is common in many environments, with at least 50 species and 70 serogroups identified. *Legionella* is commonly found in aquatic habitats where its ability to survive and to multiply within different protozoa equips the bacterium to be transmissible and pathogenic to humans.

ARCHAEA

Originally, Archaea were once thought of as extremophiles existing only in hostile environments but have since been found in all habitats and may contribute up to 20% of total biomass. Archaea are particularly common in the oceans, and the archaea in plankton may be one of the most abundant groups of organisms on the planet. Archaea are subdivided into four phyla of which two, the Crenarchaeota and the Euryarchaeota, are most intensively studied.

LACTOBACILLUS

Lactobacillus species are found in the environment mainly associated with plant material. They are also found in the gastrointestinal tract of humans, where they are symbiotic and make up a portion of the gut flora.

WINOGRADSKY

In the late 1800s and early 1900s, Sergei Winogradsky, Russian microbiologist, pioneered the investigation of microbial auto trophy, and initiated the field of Environmental Microbiology. He was a strong supporter of examining freshly-isolated organisms rather than 'domesticated' laboratory strains.

One of the methods he developed for the study of microbial nutrient cycling in the environment is what is now known as the "Windogradsky column". These can be set up in an amazing variety of ways to study sulfur, nitrogen, carbon, phosphorus, or other nutrients, most often cycling between the upper aerobic zone and the lower anaerobic zone.

DEVELOPMENTS IN BIOTECHNOLOGY

The impact of biotechnology affects the accessibility of germplasm. There has been an active interest in the use of biotechnology for germplasm work for several years. *In-vitro* collecting of germplasm and its conservation under slow growth conditions are proving to be useful additional technologies for genetic resources work. Cryopreservation techniques are also under active investigation.

Analyses of species relationships in genepools using molecular genetic techniques have provided useful results to guide collecting and utilisation programmes. The development of reliable and sensitive disease indexing procedures using antibody techniques is helping the safe transfer of germplasm throughout the world.

Thus, it is necessary to ensure that new techniques are adapted for wider application in plant genetic resources work. A pragmatic approach is required alongside more conventional technologies.

GENERAL CONSIDERATIONS

The international perspective on PGR activities, as these are taking shape today, is highlighted above. The linkages and mandates of different organisations have been reflected. It is quite evident that the problems relating to management, use and conservation of plant genetic resources are of diverse nature - scientific, technical, legal and political.

Safeguarding of crop germplasm assumes high priority. Much collections have been made and stored in genebanks. But inside agricultural experimental stations and genebanks genetic erosion continues at an alarming rate. This is particularly so in tropical countries where high temperature and humidity shorten the period during which seeds remain viable and where a considerable number of accessions can only be maintained in rather unsafe field genebanks at a high cost.

Preservation is often handicapped by lack of adequate cold storage facilities, meagre support for genebank operations, shortage of trained people or discontinuity of their service and poor financial or other incentives. These situations, which hamper genetic conservation, need to be improved.

There are, in IBPGR's network, around 140 base collections of crops and 50 centres/genebanks. This network now has spread to China, and links have been developed with the USSR. However, much needs to be done by major genebanks in comparing holdings, culling redundancy and filling in gaps. The goal should be to promote judicious exchange and consolidation among collections so that duplicate storage can be provided for every valid accession.

The quality of seeds entering storage, the efficacy of storage facilities, the extent of scientific activities, the level of management, and the security of conserved stocks of seeds vary considerably from one genebank to another. It is not surprising, therefore, that genetic erosion occurs in every genebank to some extent. A collection may suffer great losses, if it is improperly monitored for viability or if rejuvenation to produce fresh seeds is delayed.

Though, during the last two decades, the IARCs and the IBPGR have facilitated international exchange of germplasm, the scenario has changed with private seed firms/multinationals coming into picture and, in some cases, monopolising specific germplasm resources to their advantage. This has led to some apprehensions that breeding of elite cultivars might fall largely into a private sector which also produce agricultural chemicals.

Further, widespread government regulations related to the Plant Breeders' Rights could restrict the farmers from use of patented seeds and plants. This has been the subject of debate for quite sometime and recently, the Seed Association of India had organised an international symposium.

The IBPGR concern was highlighted Seed Association of India. The FAO Commission is also much involved as indicated above in these issues related to Plant Breeders' Rights and Farmers' Rights.

To mitigate the apprehensions of the developing countries, some of the activists/NGOs have widely publicised these issues which are still being debated. Possibly, steps such as the implementation of International Undertaking on Plant Genetic Resources will provide a bridge to ensure free access to crop germplasm on a global basis.

Thus, much public awareness has been generated. The society is well aware now about the importance of germplasm and its use. Practical management and policy measures need to be worked out. The outcome of the second meeting in the Dialogue Series on PGR, held in Madras in early 90s.

The above issues apart, with all our achievements on this front, there is still *a* need to augment genetic diversity from protected and natural habitats, such as unimproved germplasm grown in traditional farmlands, orchards and forestlands. The national programmes are playing a more concerted role in this direction than ever before though these are frequently handicapped by funding aspects. These gene-rich countries would need an international umbrella to support their activities.

The IBPGR has been providing the required impetus in this direction through training programmes, distribution of information, guideline manuals, free advisory service and need-based support since its inception in collecting, conservation and documentation activities;

though lately, it has taken up more of a catalyzing and advisory role. Certainly, national programmes, which are the backbone to the future build up and survival of this networking and coordination, must be strengthened further.

Concern on plant genetic resources has attained global, regional and national importance, particularly in the last two decades. The International Agricultural Research Centres (IARCs) are playing great role in the improvement and utilisation of their mandate crops, and in the collection and conservation of the concerned crop genetic diversity.

However, the overall key role in promoting concern on plant genetic resources activities in a global context and in coordinating these activities is being performed by the International Board for Plant Genetic Resources (IBPGR) in close interaction with the FAO and IARCs, and the national programmes.

The FAO Commission on Plant Genetic Resources, as an inter-governmental body, is concerned with several policy issues to ensure the efficiency of operating a global system in full coordination with national, regional and international organisations.

The role of other international organisations, such as the International Union for Conservation of Nature and Natural Resources (IUCN), the World Wide Fund for Nature (WWF) and several others has also been stressed.

Chapter 4

Acid Rains and its Harmful Effects

The term acid rain refers to what scientists call acid deposition. It is caused by airborne acidic pollutants and has highly destructive results. Scientists first discovered acid rain in 1852, when the English chemist Robert Agnus invented the term. From then until now, acid rain has been an issue of intense debate among scientists and policy makers.

Acid rain, one of the most important environmental problems of all, cannot be seen. The invisible gases that cause acid rain usually come from automobiles or coal-burning power plants. Acid rain moves easily, affecting locations far beyond those that let out the pollution. As a result, this global pollution issue causes great debates between countries that fight over polluting each other's environments.

For years, science studied the true causes of acid rain. Some scientists concluded that human production was primarily responsible, while others cited natural causes as well. Recently, more intensive research has been done so that countries have the information they need to prevent acid rain and its dangerous effects.

The levels of acid rain vary from region to region. In Third World nations without pollution restrictions, acid rain tends to be very high. In Eastern Europe, China, and the Soviet Union, acid rain levels have also risen greatly. However, because acid rain can move about so easily, the problem is definitely a global one.

For many years, there was considerable debate and disagreement over what caused acid rain. Recent scientific work, however, has helped to clarify this. The primary causes of acid rain are sulfur dioxide and nitrogen oxides.

These chemicals are released by certain industrial processes, and as a result, the more industrialized nations of Europe as well as the US suffer severely from acid rain. Most sulfur dioxide comes from power plants that use coal as their fuel. These plants emit 100 million tons of sulfur dioxide, 70% of that in the world.

Automobiles produce about half of the world's nitrogen oxide.

As the number of automobiles in use increases, so does the amount of acid rain. Power plants that burn fossil fuels also contribute significantly to nitrogen oxide emission. Though human causes are primarily responsible for acid rain, natural causes exist as well.

Fires, volcanic eruptions, bacterial decomposition, and lightening also greatly increase the amount of nitrogen oxide on the planet. However, even the gigantic explosion of Mt. St. Helens released only about what one coal power plant emits in a year. Once the tiny pollutant molecules have entered the atmosphere, they can travel for thousands of miles. Eventually, the particles will combine with other compounds to produce new, often harmful, chemicals.

Acid rain comes down to the earth in the form of rain, snow, hail, fog, frost, or dew. Once it reaches the ground, the acidity in the substance can harm and even destroy both natural ecosystems and man-made products, such as car finishes. Acid rain is having harmful effects both on people and on the natural ecosystems of the world.

Scientists today are convinced that acid rain is severe in many areas, and that it is having an adverse effect on the environments of those locations. The problem of acid rain is rapidly spreading. Because it is mainly caused by industrial processes, automobiles, and power plants, those countries that are developed have the most severe acid rain problems. However, as the undeveloped nations begin to industrialize, acid rain will increase greatly.

Determining just how much the planet is being hurt by acid rain is very difficult because the ecosystems that it affects are so diverse and complex. Many ecosystems are affected by acid rain. Bodies of water, such as lakes and rivers, see many of their inhabitants die off due to rising acidity levels. Acidic water also ruins plant nutrients, hurting plants' ability to survive and to give life to other organisms. Human-made products are also experiencing degradation from acid rain. Cars can lose their finishes, and outdoor statues are beginning to rust.

Acid rain's effects are destructive and long lasting. Though scientists have studied lakes, streams, and many other natural ecosystems to prove its negative effects, acid rain continues to be produced and is increasing in many parts of the world. Modern science has proven that acid rain is a dangerous and highly destructive problem.

As a result, various ways to limit acid rain have been invented, and some are now being used. Debate over acid rain and ways of preventing it have raged between environmentalists and corporations. Businesses such as power companies and car makers oppose controlling acid rain because they fear the effects on their profits. But in some cases, industries have attempted to curb acid rain production. The Northern States Power company began working to reduce acid rain in the 1980s, and has invested over a billion dollars to that end.

There are many ways that power plant companies like Northern States can reduce acid rain creation. They can use coal with a low sulfur content, they can remove the sulfur from smoke their plants release, and they can limit processes known to generate high levels of acid rain. Policy makers and environmental experts are now looking into the best methods to limit acid rain.

Environmentalists advocate the installation of sulfur cleaning scrubbers in factories, washing sulfur out of coal, and finding new methods of burning coal. Power plant operators are looking for less expensive solutions to the problem. Individuals can help by conserving energy or driving their cars less. Governments can pass laws restricting pollution levels, or can use a variety of methods such as tradable emission permits to reduce acid rain. Whatever way it is done, acid rain will certainly have to be limited in the future.

ACID RAIN

Acid rain is precipitation containing higher than normal amounts of nitric and sulfuric acid. According to the EPA, acid rain can either occur by wet deposition or dry deposition, occurring when sulfur dioxide and nitrogen oxide react with the atmosphere. During wet deposition the gases react with water, oxygen and other chemicals. When this happens, various acidic compounds form, most commonly sulfuric and nitric acid.

These acids are blown by the wind into areas with wet weather, where they fall in the form of rain or snow. If the acids are blown to areas with dry weather, they fall as dry deposition, clinging to dust or smoke, sticking to buildings and trees. The acid is then washed into the water system by rain.

Acid rain forms from both natural and man-made sources. Volcanoes and decaying vegetation can form sulfur dioxide and nitrogen oxides. Emissions from burning fossil fuels also produce acid rain. According to the EPA, in the United States most of the sulfur dioxide and a quarter of nitrogen oxides "come from electric power generation that relies on burning fossil fuels, like coal." Winds have been known to carry the acid rain into parts of Canada.

Measuring Acid Rain

Acid rain is measured using a pH scale.

- Pure water – pH 7. 0
- Normal rain, containing some carbon dioxide mix – pH 5. 6
- Acid rain – pH 4. 3

Acid Rain Effect on Plants and Animals

Acid rain has a negative effect on plants and animals. Once the acid gets into the water cycle it can cause the acidification of lakes and streams. The National Surface Water Survey (NSWS) identified over a thousand lakes and thousands of streams in the United States, where some form of acidification has taken place. Of these lakes and streams, 75 per cent of the acidity in the lakes was directly due to acid rain. 50 per cent of the streams had been affected by acid rain. Many areas of the Northeastern United States and Eastern Canada contain lakes and streams affected by acidification from acid rain.

Changes in the pH of lakes and streams affected by acid rain can result in a decrease in the variety of fish, plants and animals living in or near the water. Some animals and plants cannot tolerate the higher levels of acid. Snails, clams and bass are examples of animals that can only tolerate a small increase in acidity.

Acid rain destroys trees and plants by causing damage to leaves and dissolving nutrients in the surrounding soil. Trees in higher elevations, have a greater risk for damage due to acid rain because they have greater exposure to clouds carrying sulfur dioxide and nitrogen oxide.

Acid Rain Effect on Humans

Acid rain can also have a devastating effect on man-made structures, such as those made of stone and metal. Bronze statutes and marble monuments are deteriorated by acid rain. Costly repairs and maintainence is required to clean acidic compounds resulting from dry deposition collecting on buildings.

One important misconception to clear up about acid rain is its direct harm to humans. Acid rain falls from the sky just like regular rain, without odor or taste. Acid rain is not directly harmful to humans. Yet, sulfur dioxide and nitrogen oxide are harmful pollutants before they combine with water and oxygen to form acid rain. These gases cause harmful particles that can be inhaled by humans, causing lung and heart disorders.

Acid rain's effect on Plants and Wildlife

Almost everybody has heard about acid rain and knows that it is something bad. But what exactly is it? What are its effects on plants, animals, human beings, and what can be done to solve this problem? The term acid rain does not convey the true nature of the problem and therefore scientists use the term "acid depositions". This is because the acid which has formed due to pollution may return to the earth as a solid or a gas and not just as rain. Depending upon the climatic conditions it could also come down as rain, fog, or snow, and in the wet form it is known as "acid precipitation".

Sources

Certain industries, as well as emissions from vehicles give rise to increase of sulphur dioxide and nitrogen oxides in the air. These emissions change into sulphates and nitrates under the influence of sunlight and moisture, and get converted into sulphuric acid and nitric acid, which come down as acid rain.

Coal generally contains between 2 - 3% sulphur, and when it is burned, this sulphur is released into the atmosphere. Electric companies and other industries which burn coal cause a lot of emission of sulphur dioxide. Other industries which process raw ore containing sulphides in order to obtain copper, zinc, or nickel also cause an increase in sulphur dioxide levels in the atmosphere.

The major source of emissions of nitrogen oxides into the air, is from vehicles and other places where fossil fuels are burned. Forest fires, often caused by man, either deliberately or accidentally, are another source of pollution. Naturally occurring phenomena like volcanic activity, lightning, or organic decay, also gives rise to an increase in atmospheric pollutants but not much can be done about these causes. However over 90% of sulphur dioxide emissions and around 95% of nitrogen oxides released into the air are from man made sources.

The problem of acid rain is not new. It was first noticed during the 17th. century, when people observed the effects of industrialisation on plants and animals. As far back as in 1872, the Scottish chemist, Angus Robert Smith wrote a book "Air and Rain: The Beginnings of Chemical Climatology", in which he used the term "Acid Rain", and the name has stuck. The problem has become severe since the 1960s when fishermen noticed a sharp reduction in the quantity of fish in lakes of North America and Europe.

The havoc done by acid rain is not localised in the place where it is caused. The atmospheric emissions may travel for several days and over long distances depending upon wind and climatic conditions, before coming down as acid rain. The problem caused in an industrialised area may therefore result in acid rain in the surrounding forests or lakes, or even further away. It is believed that around 50% of the acid rain that occurs in Canada is due to pollution caused in the United States of America, and the effect of polluting industries in England can be felt in Norway.

If there were no pollution, the rain would still be acidic. Natural rainfall has a pH of around 6. 0. This is because of the effect of Carbon dioxide in the air which combines with water to form carbonic acid. The effect of this is however negligible, as it is neutralised in the soil by alkaline material like limestone.

However the other emissions cause the pH of the rain water to drop below 5. 5 and at this level it is considered to be acid rain. The soil cannot now neutralise the acidity of the rain water. In some places the acidification is so severe that the pH drops to around 4. 0. Rare cases have been reported of acid rain having pH of around 2 - 2. 5.

Effects of Acid Rain on Plant Life

Both natural vegetation and crops are affected by acid rain. The roots are damaged by acidic rainfall, causing the growth of the plant to be stunted, or even in its death. Nutrients present in the soil, are destroyed by the acidity. Useful micro organisms which release nutrients from decaying organic matter, into the soil are killed off, resulting in less nutrients being available for the plants.

The acid rain, falling on the plants damages the waxy layer on the leaves and makes the plant vulnerable to diseases. The cumulative effect means that even if the plant survives it will be very weak and unable to survive climatic conditions like strong winds, heavy rainfall, or a short dry period. Plant germination and reproduction is also inhibited by the effects of acid rain.

Effects of Acid Rain on Aquatic Life

The action of acid rain causes harmful elements like mercury and aluminium to be leached from the soil and rocks and it is then carried into the lakes where aquatic life may be affected. Warning signs have been posted at several lakes, telling about the dangers of eating fish which may have been poisoned by mercury. Just as the soil has a natural ability to neutralise the acidity of rain water, within a certain limit, so also lakes and other water bodies can to a certain extent nullify the effects of acid rain.

However as the acidity increases, the natural mechanisms are no longer able to cope. As the water gets more acidic its pH goes down. As the pH reaches 5. 5, plankton, certain insects and crustaceans begin to die. At a pH of around 5. 0, the fish population begin to die. When the pH drops below 5. 0, all the fish have died, and the bottom of the lake lies covered with undecayed material. Every year during the spring thaw, there is a sudden increase in the acidity of the lakes as frozen acid is suddenly deposited in them. This "Acid Shock" prevents the reproduction of aquatic species, or results in the deaths of the hatchlings.

Effects on Animals and Birds

All living organisms are interdependent on each other. If a lower life form is killed, other species that depended on it will also be affected. Every animal up the food chain will be affected. Animals and birds, like waterfowl or beavers, which depended on the water for food sources or as a habitat, also begin to die. Due to the effects of acid rain, animals which depended on plants for their food also begin to suffer. Tree dwelling birds and animals also begin to languish due to loss of habitat.

Effects on Human Beings

Mankind depends upon plants and animals for food. Due to acid rain the entire fish stocks in certain lakes have been wiped out. The economic livelihood of people who depended on fish and other aquatic life suffers as a result. Eating fish which may have been contaminated by mercury can cause serious health problems. In addition to loss of plant and animal life as food sources, acid rain gets into the food we eat, the water we drink, as well as the air we breathe.

Due to this asthmatic people and children are directly affected. Urban drinking water supplies are generally treated to neutralise some of the effects of acid rain and therefore city dwellers may not directly suffer due to acidified drinking water. But out in the rural areas, those depending upon lakes, rivers, and wells will feel the effects of acid rain on their health.

The acidic water moving through pipes causes harmful elements like lead and copper to be leached into the water. Aluminium which dissolves more easily in acid rain as compared to pure rainfall, has been linked to Alzheimer's disease. The treatment of urban water supplies may not include removal of elements like Aluminium, and so is a serious problem in cities too.

Other Effects

All living things, whether plants or animals, whether living on land or in the water or trees, are affected either directly or indirectly by acid rain. Even buildings, bridges and other structures are affected.

In cities, paint from buildings have peeled off and colours of cars have faded due to the effects of acid rain.

From the Taj Mahal in India to the Washington Monument great buildings all over the world have been affected by the acid rainfall which causes corrosion, fracturing, and discoloration in the structures. In Europe, structures like The Acropolis in Greece and Renaissance buildings in Italy, as well as several churches and cathedrals have suffered visible damage.

In the Yucatan peninsula in Mexico, and in places in South America, ancient Mayan Pyramids are being destroyed by the acid rain. Temples, murals, and ancient inscriptions which had previously survived for centuries are now showing severe signs of corrosion.

Even books, manuscripts, paintings, and sculpture are being affected in museums and libraries, where the ventilation system cannot eliminate the acid particles from the air which circulates in the building. In some parts of Poland, trains are required to run slowly, as the tracks are badly damaged due to corrosion caused by acid rainfall.

Solutions of Acid Rain

The bottom line is that all things on earth are being affected by this problem and the good news is that something is being done to solve it. Pressure from the environmental groups, and public has increased as the effects of the havoc caused by acid rain become more apparent. Governments all over the world have drawn up plans to tackle this problem.

Lakes that have become highly acidic, can be treated by adding large quantities of alkaline substances like quicklime, in a process called liming. Although it has worked in several places, it has not been successful where the lake is very large, making this procedure economically unfeasible, or in other lakes where the flushing rate of the lake waters is too large resulting in the lake becoming acidic again.

The best approach seems to be in prevention. To this end environmental regulations have been enacted to limit the quantity of emissions released in the atmosphere. Several industries have added scrubbers to their smoke stacks to reduce the amount of sulphur dioxide dumped in the atmosphere.

Specially designed catalytic converters are used to ensure that the gases coming out from exhaust pipes of automobiles, are rendered harmless. Several industries which use coal as fuel have begun to wash the coal before using it thereby reducing the amount of Sulphur present in it, and consequently the amount of emissions. Usage of coal with a low Sulphur content also reduces the problem.

We as individuals can take several steps to alleviate the effects of this problem. A reduction in use of vehicles will reduce the amount of emission caused by our vehicles. So do not use the car unless it is absolutely required. For going short distances, walk or try to use a bicycle.

This will not only protect the environment but also improve your health. If the distance is greater, try using public transportation. If you must use your vehicle try forming a car pool and share your vehicle with someone else. Ensure that your vehicle is properly tuned, and fitted with a catalytic converter, to reduce the emissions.

Reduce use of electric power. Switch off lights, and other electrical appliances when not required. Do not leave your Televisions, VCRs, Microwave Ovens or Music Systems on Stand-by when not required. Switch them off.

Reducing power consumption will reduce the amount of coal burnt to produce electricity, and thus reduce the amount of pollution. This is true even if your electricity company does not use coal for producing electricity, but some other more environmentally friendly way. This is because the electricity you have saved can now be used elsewhere, thus benefiting nature.

Speak to others about this problem. Increasing awareness is one way of ensuring that things are done to solve this global problem. Find out what fuel is being used by your electricity company to produce electricity. If they use coal, ask what methods they use to contain, if not eliminate, the problem of sulphur emissions. Washing the coal used, or using coal having a low sulphur content, is costly and therefore some companies try to avoid this. If you have the option, switch to a utility that shows more concern for the environment.

CHEMISTRY OF ACID RAIN

Acid Rain Normally while rain travels through the air, it dissolves floating chemicals and washes down particles that are suspended in air. At the start of its journey raindrops are neutral (pH = 7). In clean air, rain picks up materials that occur naturally such as dust, pollen, some CO_2 and other chemicals produced by lightening or volcanic activities.

These substances make rain slightly acidic (pH = 6), which is not dangerous. However, when rain falls through polluted air, it comes across chemicals such as gaseous oxides of sulphur (SO_x), oxides of nitrogen (NO_x), mists of acids such as hydrochloric and phosphoric acid, released from automobile exhausts industrial plants, electric power plants etc. These substances dissolve in falling rain making it more acidic than normal with pH range between 5. 6 -3. 5. In some case, it's pH gets lowered to the extent of 2. This leads to acid rain. The term acid rain is used here to describe all types of precipitation, namely, rain, snow, fog and dew more acidic than normal.

Chemistry of acid Rain

In the natural processes of volcanic eruptions, forest fires and bacterial decomposition of organic oxides of sulphur and nitrogen, production and reductions of gases naturally tend to an equilibrium.

Power plants, smelting plants, industrial plants, burning of coal and automobile exhausts, release additional sulphur dioxide, nitrogen oxides and acidic soot, causing pollution. Sulphur dioxide and nitrogen dioxide interact with water vapours in presence of sunlight to form sulphuric acid and nitric acid mist.

$$SO_2 + H_2O \longrightarrow \underset{\text{Sulphurous Acid}}{H_2SO_3}$$

$$\underset{\text{Ozone}}{SO_2 + O_3} \longrightarrow SO_3 + O_2 ; SO_3 + H_2O \longrightarrow \underset{\text{Sulphuric acid}}{H_2SO_4}$$

$$2NO + O_2 \longrightarrow 2NO_2 ; 2NO_2 + H_2O \longrightarrow \underset{\text{Nitric acid}}{HNO_3} + \underset{\text{Nitrous acid}}{HNO_2}$$

The formed sulphuric acid and nitric acid remain as vapour at high temperatures. These begin to condense as the temperature falls and mix with rain or snow, on the way down to the Earth and make rain sufficiently acidic.

HARMFUL EFFECTS OF ACID RAIN

SO_x, NO_x mixed with water as acid rain causes plant, animal and material damage. Some of the significant ill effects of acid rain are:

Damage to Animals

Acid rain chemically strips waterways of necessary nutrients and lowers the pH to levels where plants and animals cannot live. Most of the aquatic animals cannot survive when the pH is less than 4. Some species of fish, such as salmon, die even when the pH is less than 5. 5. Certain species of algae and zooplankton are eliminated at pH less than 6. A reduction in the zooplankton and bottom fauna ultimately affects the food availability for the fish population. The problem is most severe downwind of industrial areas where fishing and tourism are major sources of income such as in Norway and Sweden.

Damage to Plants

Acidic water is dangerous to plants. Sulphuric and nitric acid rain washes nutrients out of the soil, damages the bark and leaves of trees and harms the fine root hairs of many plants which are needed to absorb water. Leaf pigments are decolorized because acid affects green pigment (chlorophyll) of plants.

Agricultural productivity is also decreased. Several non-woody plants, such as barley, cotton and fruit trees like apple, pear, etc. , are severely affected by acid rain. Since the acid concentration increases near the base of clouds by density, high altitude trees and vegetation may be exposed to pH levels as low as 3. Unique areas such as the Black Forest in Germany and sugar maples in Vermont (USA) are particularly threatened.

Material Damage

Metallic surfaces exposed to acid rain are easily corroded. Textile fabrics, paper and leather products lose their material strength or disintegrate by acid rain. Building materials such as limestone, marble, dolomite, mortar and slate are weakened on reaction with acid rains because of the formation of soluble compounds.

$$CaCO_3 + H_2SO_4 \longrightarrow CaSO_4 + H_2O + CO_2$$

Thus, acid rain is dangerous for historical monuments.

Chapter 5

Importance of Degradation

RECYCLING WASTES INTO VALUABLE ORGANIC FERTILIZER

Environmental degradation is a major threat confronting the world, and the rampant use of chemical fertilizers contributes largely to the deterioration of the environment through depletion of fossil fuels, generation of carbon dioxide (CO_2) and contamination of water resources. It leads to loss of soil fertility due to imbalanced use of fertilizers that has adversely impacted agricultural productivity and causes soil degradation.

Now there is a growing realization that the adoption of ecological and sustainable farming practices can only reverse the declining trend in the global productivity and environment protection. On one hand tropical soils are deficient in all necessary plant nutrients and on the other hand large quantities of such nutrients contained in domestic wastes and agricultural byproducts are wasted. It is estimated that in cities and rural areas of India nearly 700 million t organic waste is generated annually which is either burned or land filled.

Such large quantities of organic wastes generated also pose a problem for safe disposal. Most of these organic residues are burned currently or used as land fillings. In nature's laboratory there are a number of organisms (micro and macro) that have the ability to convert organic waste into valuable resources containing plant nutrients and organic matter, which are critical for maintaining soil productivity. Microorganisms and earthworms are important biological organisms helping nature to maintain nutrient flows from one system to another and also minimize environmental degradation.

The earthworm population is about 8–10 times higher in uncultivated area. This clearly indicates that earthworm population decreases with soil degradation and thus can be used as a sensitive indicator of soil degradation.

In this report a simple biotechnological process, which could provide a 'win-win' solution to tackle the problem of safe disposal of waste as well as the most needed plant nutrients for sustainable productivity is described.

VERMICOMPOSTING

Vermicomposting is a simple biotechnological process of composting, in which certain species of earthworms are used to enhance the process of waste conversion and produce a better end product. Vermicomposting differs from composting in several ways. It is a mesophilic process, utilizing microorganisms and earthworms that are active at 10–32°C (not ambient temperature but temperature within the pile of moist organic material).

The process is faster than composting; because the material passes through the earthworm gut, a significant but not yet fully understood transformation takes place, whereby the resulting earthworm castings (worm manure) are rich in microbial activity and plant growth regulators, and fortified with pest repellence attributes as well! In short, earthworms, through a type of biological alchemy, are capable of transforming garbage into 'gold'.

Importance of Vermicompost Source of Plant Nutrients

Earthworms consume various organic wastes and reduce the volume by 40–60%. Each earthworm weighs about 0. 5 to 0. 6 g, eats waste equivalent to its body weight and produces cast equivalent to about 50% of the waste it consumes in a day. These worm castings have been analyzed for chemical and biological properties. The moisture content of castings ranges between 32 and 66% and the pH is around 7. 0. The worm castings contain higher per centage (nearly twofold) of both macro and micronutrients than the garden compost.

Table. Nutrient Composition of Vermicompost and Garden Compost.

Nutrient element	Vermicompost (%)	Garden compost (%)
Organic carbon	9. 8-13. 4	12. 2
Nitrogen	0. 51-1. 61	0. 8
Phosphorus	0. 19-1. 02	0. 35
Potassium	0. 15-0. 73	0. 48
Calcium	1. 18-7. 61	2. 27
Magnesium	0. 093-0. 568	0. 57
Sodium	0. 058-0. 158	<0. 01
Zinc	0. 0042-0. 110	0. 0012
Copper	0. 0026-0. 0048	0. 0017
Iron	0. 2050-1. 3313	1. 1690
Manganese	0. 0105-0. 2038	0. 0414

From earlier studies also it is evident that vermicompost provides all nutrients in readily available form and also enhances uptake of nutrients by plants. Sreenivas et al. studied the integrated effect of application of fertilizer and vermicompost on soil available nitrozen (N) and uptake of ridge gourd (*Luffa acutangula*) at Rajendranagar, Andhra Pradesh, India.

Soil available N increased significantly with increasing levels of vermicompost and highest N uptake was obtained at 50% of the recommended fertilizer rate plus 10 t ha^{-1} vermicompost. Similarly, the uptake of N, phosphorus (P), potassium (K) and magnesium (Mg) by rice (*Oryza sativa*) plant was highest when fertilizer was applied in combination with vermicompost.

Plant Growth Promoting Activity

Growth promoting activity of vermicompost was tested using a plant bioassay method. The plumule length of maize (*Zea mays*) seedling was measured 48 h after soaking in vermicompost water and in normal water. The marked difference in plumule length of maize seedlings indicated that plant growth promoting hormones are present in vermicompost.

Table. Plumule Length of Maize Seedings

Treatment	Initial Length(cm)	Final Length (cm)
Tank Water	16. 5	16. 6
Vermicompost Water	17. 6	18. 6

Improved Crop Growth and Yield

Vermicompost plays a major role in improving growth and yield of different field crops, vegetables, flower and fruit crops. The application of vermicompost gave higher germination (93%) of mung bean (*Vigna radiata*) compared to the control (84%).

Further, the growth and yield of mung bean was also significantly higher with vermicompost application. Likewise, in another pot experiment, the fresh and dry matter yields of cowpea (*Vigna unguiculata*) were higher when soil was amended with vermicompost than with biodigested slurry.

The efficiency of vermicompost was evaluated in a field study by Desai et al. They stated that the application of vermicompost along with fertilizer N gave higher dry matter (16. 2 g $plant^{-1}$) and grain yield (3. 6 t ha^{-1}) of wheat (*Triticum aestivum*) and higher dry matter yield (0. 66 g $plant^{-1}$) of the following coriander (*Coriandrum sativum*) crop in sequential cropping system. Similarly, a positive response was obtained with the application of vermicompost to other field crops such as sorghum (*Sorghum bicolor*) (Patil and Sheelavantar 2000) and sunflower (*Helianthus annuus*). Application of vermicompost at 5 t ha^{-1} significantly increased yield of tomato (*Lycopersicon esculentum*) (5. 8 t ha^{-1}) in farmers' fields in Adarsha watershed, Kothapally, Andhra Pradesh compared to control (3. 5 t ha^{-1}).

Similarly, greenhouse studies at Ohio State University in Columbus, Ohio, USA have indicated that vermicompost enhances transplant growth rate of vegetables. Amendment of vermicompost with a transplant grown without vermicompost had the highest amount of red marketable fruit at harvest.

In addition, there were no symptoms of early blight lesions on the fruit at harvest. The yield of pea (*Pisum sativum*) was also higher with the application of vermicompost (10 t ha^{-1}) along with recommended N, P and K than with these fertilizers alone. Vadiraj et al. reported that application of vermicompost produced herbage yields of coriander cultivars that were comparable to those obtained with chemical fertilizers.

The fresh weight of flowers such as *Chrysanthemum chinensis* increased with the application of different levels of vermicompost. Also, the number of flowers per plant, flower diameter (6 cm) and yield (0. 5 t ha^{-1}) were maximum with the application of 10 t ha^{-1} of vermicompost along with 50% of recommended dose of NPK fertilizer. However, the vase life of flowers (11 days) was high with the combined application of vermicompost at 15 t ha^{-1} and 50% of recommended dose of NPK fertilizer.

Reduction in Soil C:N Ratio

Vermicomposting converts household waste into compost within 30 days, reduces the C:N ratio and retains more N than the traditional methods of preparing composts. The C:N ratio of the unprocessed olive cake, vermicomposted olive cake and manure were 42, 29 and 11, respectively. Both the unprocessed olive cake and vermicomposted olive cake immobilized soil N throughout the study duration of 91 days.

Cattle manure mineralized an appreciable amount of N during the study. The prolonged immobilization of soil N by the vermicomposted olive cake was attributed to the C:N ratio of 29 and to the recalcitrant nature of its C and N composition. The results suggest that for use of vermicomposted dry olive cake as an organic soil amendment, the management of vermicomposting process should be so adjusted as to ensure more favorable N mineralizationimmobilization.

Role in Nitrogen Cycle

Earthworms play an important role in the recycling of N in different agroecosystems, especially under *jhum* (shifting cultivation) where the use of agrochemicals is minimal. Bhadauria and Ramakrishnan reported that during the fallow period intervening between two crops at the same site in 5- to 15-year *jhum* system, earthworms participated in N cycle through cast-egestion, mucus production and dead tissue decomposition.

Soil N losses were more pronounced over a period of 15-year *jhum* system. The total soil N made available for plant uptake was higher than the total input of N to the soil through the addition of slashed vegetation, inorganic and organic manure, recycled crop residues and weeds.

Improved Soil Physical, Chemical and Biological Properties

Limited studies on vermicompost indicate that it increases macropore space ranging from 50 to 500 ìm, resulting in improved air-water relationship in the soil which favorably affect plant growth. The application of organic matter including vermicompost favorably affects soil pH, microbial population and soil enzyme activities.

It also reduces the proportion of water-soluble chemical species, which cause possible environmental contamination.

TYPES OF EARTHWORMS

Earthworms are invertebrates. There are nearly 3600 types of earthworms in the world and they are mainly divided into two types:

- Burrowing;
- Non-burrowing.

The burrowing types *Pertima elongata* and *Pertima asiatica* live deep in the soil. On the other hand, the non-burrowing types *Eisenia fetida* and *Eudrilus eugenae* live in the upper layer of soil surface. The burrowing types are pale, 20 to 30 cm long and live for 15 years. The non-burrowing types are red or purple and 10 to 15 cm long but their life span is only 28 months.

The non-burrowing earthworms eat 10% soil and 90% organic waste materials; these convert the organic waste into vermicompost faster than the burrowing earthworms. They can tolerate temperatures ranging from 0 to 40°C but the regeneration capacity is more at 25 to 30°C and 40–45% moisture level in the pile.

The burrowing type of earthworms comes onto the soil surface only at night. These make holes in the soil up to a depth of 3. 5 m and produce 5. 6 kg casts by ingesting 90% soil and 10% organic waste.

Earthworm Multiplication

Numerous organic materials have been evaluated for growth and reproduction of earthworms as these materials directly affect the efficacy of vermicompost. Nogales et al. evaluated the suitability of dry olive cake, municipal biosolids and cattle manure as substrates for vermicomposting. They reported that larger weights of newly hatched earthworms were obtained in substrate containing dry olive cake.

In another study, maize straw was found to be the most suitable feed material compared to soybean (*Glycine max*) straw, wheat straw, chickpea (*Cicer arientinum*) straw and city refuse for the tropical epigeic earthworm, *Perionyx excavatus*. Zajonc and Sidor evaluated and compared various non-standard materials for the preparation of vermicompost.

A mixture of cotton waste with cattle manure in the ratio of 1:5 was found to be the best. The use of grape cake alone increased earthworm weight slightly. Tobacco (*Nicotiana tabacum*) waste, used as substrate, increased earthworm weight but the earthworms failed to reproduce. A mixture of tobacco waste with rabbit manure in the ratio of 1:5 was found to be lethal to the earthworms.

A multiplication trial was conducted at the International Crops Research Institute for the Semi-Arid Tropics (ICRISAT), Patancheru, Andhra Pradesh with three kinds of earthworm cultures (*Eisenia fetida, Eudrilus eugenae* and *Perionyx excavatus*) using wheat straw, chickpea straw, tree leaves (*Peltophorum* sp) and *Parthenium* mixed with cow dung as feed materials.

There was an increase in earthworm population and size during incubation for 90 days. The three types of earthworms multiplied 12 to 18 times when grown individually using legume tree leaves and cow dung mixture as raw material.

However, mixed culture (of all three species) showed higher multiplication rate (27 times) than the individual species. Further studies on earthworm multiplication were also conducted at ICRISAT using tree leaves and *Gliricidia* stems mixed with cattle manure as feed material. The earthworm population decreased when grown in mixture of *Gliricidia* stems and cattle manure.

These results indicated that *Gliricidia* loppings could not be used for multiplication of earthworms. *Gliricidia* bark is known to possess toxic properties as it is used as rat poisoning bait. In another multiplication study at ICRISAT, there was maximum increase in earthworm population (570%) and weight (109%) when grown in a feed material containing tree leaves (3 kg) and cow dung (6 kg).

In contrast, mortality of earthworms (about 7 to 22%) was observed by growing them in a feed material containing soil. All these studies indicated that *Gliricidia* and tobacco leaves are not suitable for multiplication of earthworms. Perhaps the alkaloids and other principal compounds present in these leaves may effect the survival of earthworms. Also, soil and rabbit manure should not be mixed with earthworm feed material.

Table. Multiplication Trial of Earthworm Species at ICRISAT, Patancheru, India in 2000.

Earthworm species	Initial population	Final population	Increase (%)
Mixed culture	900	15950	1612 (27)
Eisenia fetida	90	1036	1051 (12)
Eudrilus eugenae	55	1007	1731 (18)
Perionyx excavatus	85	1192	1302 (14)

Table. Multiplication trials of earthworms using different organic materials at ICRISAT,Patancheru, India during 2000-02.

Earthworm species	Feed material	Initial Popul-ation	Initial Weight (g)	Final[1] Popul-ation	Final[1] Weight (g)
Eisenia fetida	Tree leaves (15 kg)	345	20	2510	207
	Cattle manure (15 kg)	510	207	1159	207
	Cattle manure (3 kg) + Gliricidia stem (6 kg)	1255	101	1000	50
Eudrilus eugenae	Tree leaves (15 kg)	311	21	2986	334
	Cattle manure (15 kg)	2986	334	1522	216
	Cattle manure	2707	230	2249	100

	(3 kg) +Z Gliricidia stem (6 kg)				
Perionyx excavatus	Tree leaves (15 kg)	409	29	2707	230
	Cattle manure (15 kg)	2707	230	2650	187
	Cattle manure (3 kg) + Gliricidia stem (6 kg)	3356	365	1000	50

Table. Multiplication Trials of Mixed Culture of Earthworms Using Soil and Other Organic Substrates at ICRISAT, Patancheru, India, 2000-02.

	Initial		Final		Increase1 (%)	
Feed material	Number	Weight(g)	Number	Weight(g)	Number	Weight
Cow dung (15 kg)	500	89	750	163	50	83
Tree leaves (3 kg) + cow dung (3 kg)	500	95	1545	125	21	32
Tree leaves (3 kg) + cow dung (6 kg)	500	110	3351	230	570	109
Pigeonpea leaves + pod shells + tree leaves (2 kg) + cow dung (2 kg)	500	98	2230	187	346	90
Pigeonpea leaves + pod shells + tree leaves (2 kg) + cow dung	500	115	1490	193	198	68

(4 kg)						
Soil (5 kg) + cow dung (5 kg)	1000	90	784	87	–22	–3
Soil (5 kg) + cow dung (5 kg) + pigeonpea leaves (1 kg)	1000	75	1023	241`	2	223
Soil (5 kg) + cow dung (5 kg) + tree leaves (1 kg)	1000	160	929	170	–7	–6

Temperature Changes During the Process

Change in temperature was observed during the process of vermicomposting (from 5 to 65 days) with different farm residues (*Parthenium* and grass). In the beginning of the process, ie, up to 15 days, the temperature was high (32 to 33°C) in both *Parthenium* and grass substrates when compared to outside temperature (26 to 30°C).

Later, there was a gradual decrease in temperature, which reached a minimum of about 24°C. However, higher temperature was recorded in *Parthenium* compost (decline from 32. 8 to 27. 5°C) than in grass compost (decline from 31. 5 to 26. 8°C) during the whole period of digestion process.

Generally more heat was evolved from control treatment (without earthworms) than the vermicompost treatments (with earthworms). From these studies, it was suggested that the most suitable period for releasing the earthworms into organic residues would be between 15 and 20 days after heaping of the organic residues when the temperature is about 25°C.

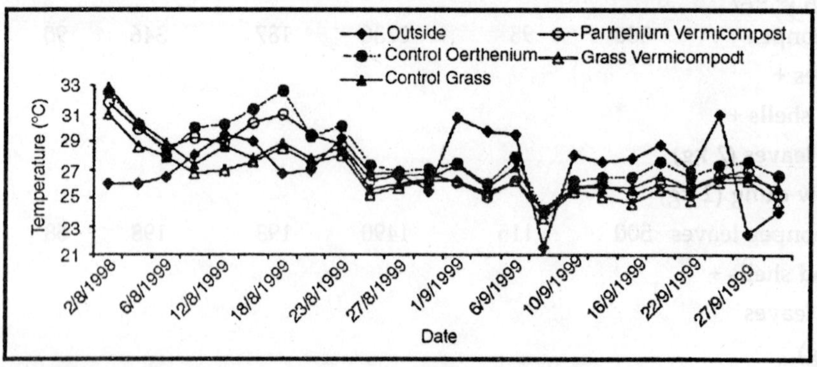

Fig. Temperature Changes During Biodigestion.

Methods of Vermicomposting Pits Below the Ground

Pits made for vermicomposting are 1 m deep and 1. 5 m wide. The length varies as required.

Heaping Above the Ground

The waste material is spread on a polythene sheet placed on the ground and then covered with cattle dung. Sunitha et al. compared the efficacy of pit and heap methods of preparing vermicompost under field conditions. Considering the biodegradation of wastes as the criterion, the heap method of preparing vermicompost was better than the pit method.

Earthworm population was high in the heap method, with a 21-fold increase in *Eudrilus eugenae* as compared to 17-fold increase in the pit method. Biomass production was also higher in the heap method (46-fold increase) than in the pit method (31-fold). Consequent production of vermicompost was also higher in the heap method (51 kg) than in the pit method (40 kg).

Tanks Above the Ground

Tanks made up of different materials such as normal bricks, hollow bricks, shabaz stones, asbestos sheets and locally available rocks were evaluated for vermicompost preparation. Tanks can be constructed with the dimensions suitable for operations.

At ICRISAT, we have evaluated tanks with dimensions of 1. 5 m (5 feet) width, 4. 5 m (15 feet) length and 0. 9 m (3 feet) height. The commercial biodigester contains a partition wall with small holes to facilitate easy movement of earthworms from one tank to the other.

CEMENT RINGS

Vermicompost can also be prepared above the ground by using cement rings. The size of the cement ring should be 90 cm in diameter and 30 cm in height.

Commercial Model

The commercial model for vermicomposting developed by ICRISAT consists of four chambers enclosed by a wall (1. 5 m width, 4. 5 m length and 0. 9 m height). The walls are made up of different materials such as normal bricks, hollow bricks, shabaz stones, asbestos sheets and locally available rocks. This model contains partition walls with small holes to facilitate easy movement of earthworms from one chamber to another. Providing an outlet at one corner of each chamber with a slight slope facilitates collection of excess water, which is reused later or used as earthworm leachate on crop. The four components of a tank are filled with plant residues one after another. The first chamber is filled layer by layer along with cow dung and then earthworms are released. Then the second chamber is filled layer by layer. Once the contents in the first chamber are processed the earthworms move to chamber 2, which is already filled and ready for earthworms.

This facilitates harvesting of decomposed material from the first chamber and also saves labour for harvesting and introducing earthworms. This technology reduces labour cost and saves water as well as time.

Fig. Commercial Model for Vermicomposting at ICRISAT.

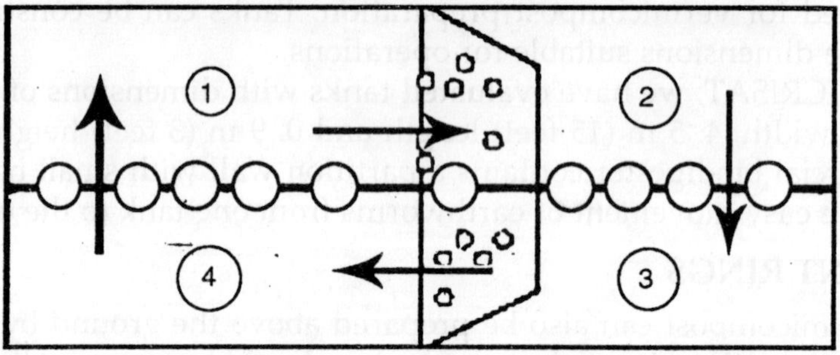

Fig. Diagrammatic Representation of the Commercial Model with four Chamber for vermicomposting

Materials Required for Vermicomposting

A range of agricultural residues, all dry wastes, for example, sorghum straw and rice straw (after feeding cattle), dry leaves of crops and trees, pigeonpea (*Cajanus cajan*) stalks, groundnut (*Arachis hypogaea*) husk, soybean residues, vegetable wastes, weed (*Parthenium*) plants before flowering, fibre from coconut (*Cocos nucifera*) trees and sugarcane (*Saccharum officinarum*) trash can be converted into vermicompost.

In addition, animal manures, dairy and poultry wastes, food

industry wastes, municipal solid wastes, biogas sludge and bagasse from sugarcane factories also serve as good raw materials for vermicomposting.

The quantity of raw materials required using a cement ring of 90 cm in diameter and 30 cm in height or a pit or tank measuring 1. 5 m × 1 m × 1 m is given below:

Dry organic wastes (DOW)	50 kg
Dung slurry (DS)	15 kg
Rock phosphate (RP)	2 kg
Earthworms (EW)	500-700
Water	5 L every three days

The various ingredients are used in the ratio of 5:1. 5:0. 2:50–75:0. 5 of DOW:DS:RP:EW:W. In the tank or pit system 100 kg of raw material and 15–20 kg of cow dung are needed for each cubic meter of the bed.

Vermicompost Preparation Steps in the Process

Vermicomposting involves the following steps which are depicted in Figure:

- Cover the bottom of the cement ring with a layer of tiles or coconut husk or polythene sheet
- Spread 15–20 cm layer of organic waste material on the polythene sheet. Sprinkle rock phosphate powder if available (it helps in improving nutritional quality of compost) on the waste material and then sprinkle cow dung slurry. Fill the ring completely in layers as described. Paste the top of the ring with soil or cow dung. Allow the material to decompose for 15 to 20 days.
- When the heat evolved during the decomposition of the materials has subsided (15–20 days after heaping), release selected earthworms (500 to 700) through the cracks developed.
- Cover the ring with wire mesh or gunny bag to prevent birds from picking the earthworms. Sprinkle water every three days to maintain adequate moisture and body temperature of the earthworms.
- The vermicompost is ready in about 2 months if agricultural waste is used and about 4 weeks if sericulture waste is used as substrate.
- The processed vermicompost is black, light in weight and free from bad odor.
- When the compost is ready, do not water for 2–3 days to make

compost easy for sifting. Pile the compost in small heaps and leave under ambient conditions for a couple of hours when all the worms move down the heap in the bed. Separate upper portion of the manure and sieve the lower portion to separate the earthworms from the manure. The culture in the bed contains different stages of the earthworm's life cycle, namely, cocoons, juveniles and adults. Transfer this culture to fresh half decomposed feed material. The excess as well as big earthworms can be used for feeding fish or poultry. Pack the compost in bags and store the bags in a cool place.

- Prepare another pile about 20 days before removing the compost and repeat the process by following the same procedure as described above.

Precautions During the Process

The following precautions should be taken during vermicomposting:

- The African species of earthworms, *Eisenia fetida* and *Eudrilus eugenae* are ideal for the preparation of vermicompost. Most Indian species are not suitable for the purpose.
- Only plant-based materials such as grass, leaves or vegetable peelings should be utilized in preparing vermicompost.
- Materials of animal origin such as eggshells, meat, bone, chicken droppings, etc are not suitable for preparing vermicompost.
- *Gliricidia* loppings and tobacco leaves are not suitable for rearing earthworms.
- The earthworms should be protected against birds, termites, ants and rats.
- Adequate moisture should be maintained during the process. Either stagnant water or lack of moisture could kill the earthworms.
- After completion of the process, the vermicompost should be removed from the bed at regular intervals and replaced by fresh waste materials.

How to Use Vermicompost

- *Vermicompost can be used for all crops*: agricultural, horticultural, ornamental and vegetables at any stage of the crop.
- *For general field crops*: Around 2–3 t ha^{-1} vermicompost is used by mixing with seed at the time of sowing or by row application when the seedlings are 12–15 cm in height. Normal irrigation is followed.

- *For fruit trees*: The amount of vermicompost ranges from 5 to 10 kg per tree depending on the age of the plant. For efficient application, a ring (15–18 cm deep) is made around the plant. A thin layer of dry cow dung and bone meal is spread along with 2–5 kg of vermicompost and water is sprayed on the surface after covering with soil.
- *For vegetables*: For raising seedlings to be transplanted, vermicompost at 1 t ha^{-1} is applied in the nursery bed. This results in healthy and vigorous seedlings. But for transplants, vermicompost at the rate of 400–500 g per plant is applied initially at the time of planting and 45 days after planting (before irrigation).
- *For flowers*: Vermicompost is applied at 750–1000 kg ha^{-1}.
- For vegetable and flower crops vermicompost is applied around the base of the plant. It is then covered with soil and watered regularly.

Vermicomposting Process.

Fig. (a) Plastic Sheet Placed Below the Ring. (b) Layer of Raw Material Placed on Polythene Sheet.

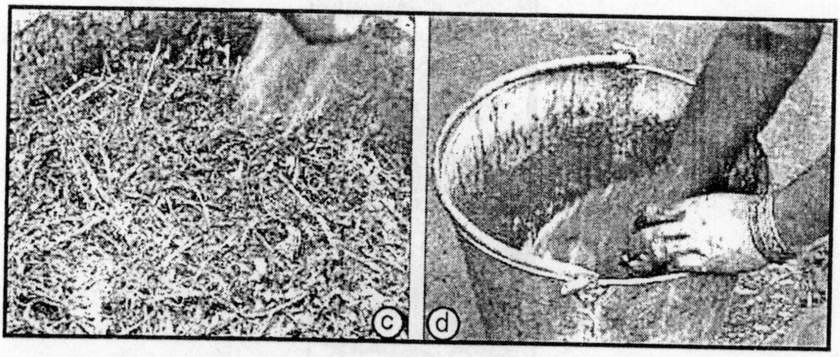

Fig. (c) Rock Phosphate Powder Sprinkled on Organic Material. (d) Cow dung Slurry.

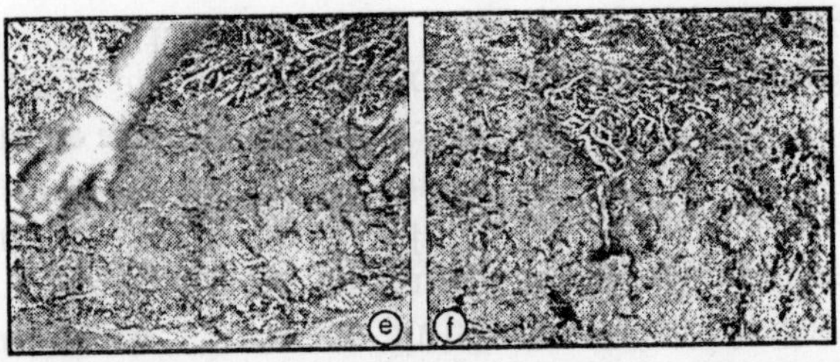

Fig. (e)Cement Ring Sealed with cow Dung. (f) Earthworms are Released Near Cracks.

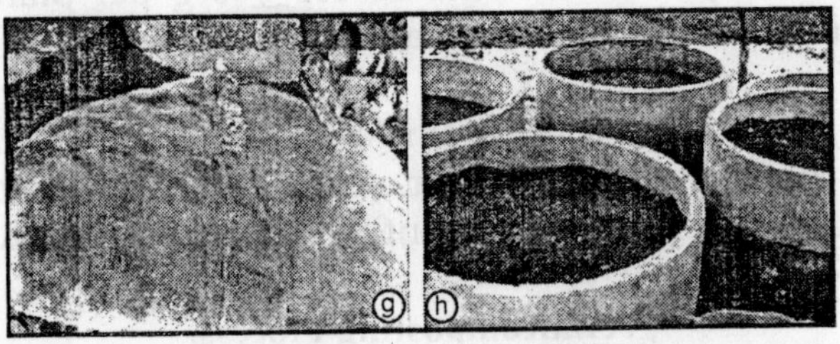

Fig. (g) Cement Ring Covered with GunnyBag. (h) Processed Vermicompost.

Fig. (i) Compost Sieved.

Fig. (j) Bag filled with Vermicompost

Biodiversity in Vermicompost

In the present study, vermicompost samples were collected and analyzed for microbial diversity and population studies. The vermicompost samples were collected in sterile containers from the rings before harvesting the compost.

To compare microbial diversity, samples from the partially decomposed dry organic waste material, ready for the release of the earthworms, were also collected and checked for diversity and population counts.

Total mircobial populations of bacteria, fungi and actinomycetes from the substrates were determined by using dilution plate techniques with suitable media (Nutrient Agar, Potato Dextrose Agar, Actinomycetes Isolation Agar-HI Media). The number of colony forming units (CFU) was expressed as CFU g^{-1}.

Several authors have noted that the earthworms play a major role in affecting populations of soil organisms, especially in causing changes in the soil microbial community. The present work recorded higher microbial populations in the partially decomposed dry organic waste material for vermicompost than the vermicompost.

This may be due to the existing temperatures and pH in the partially decomposed raw material. But compared to conventional thermophilic composts, vermicompost is much richer in microbial diversity, populations and activities.

Table. Microbital Populations from the Samples of Vermicompost.

	Bacteria (CFU $^{g-1}$)	Fungi (CFU $^{g-1}$)	Actinomycetes (CFU $^{g-1}$)
Vermicompost	54×10^6	8×10^4	1×10^4
Partially decomposed dry organic waste material for vermicompost	69×10^6	11×10^4	2×10^4

The fungal isolates from the samples were identified upto species level. Much diversity was observed between the two samples collected. *Aspergillus, Fusarium, Mucor, Cladosporium,* and *Trichoderma* were the common genera observed in both the samples. Genera like *Absidia,* and *Stachbotrys* were recorded in vermicompost.

Genera like *Alternaria, penicillium,* and *Thermomyces* were isolated from partially decomposed dry organic waste material for vermicompost. This clearly indicates that the fungal diversity is more in the decomposed material than in the vermicompost. The digestive epithelium of the simple straight tubular gut of worms is known to secrete cellulase, amylase, invertase, protease, phosphatase.

Earthworms inevitably consume the soil microbes during the ingestion of litter and soil. It has been recently estimated that earthworms necessarily have to feed on microbes, particularly fungi for their protein/nitrogen requirement. This may be the reason for the less diversity of fungi and microbial counts seen in the vermicompost collected.

In both the samples per centage of *Aspergillus* was more when compared with other genera. *Tricoderma* and *Penicillium* have antibiotic activities and can also be used as biological control on soil borne pathogens. Only a few studies have investigated that the suppression of soil borne plant pathogens by vermicompost, or disease suppression in the presence of earthworms.

Disease suppression by compost has been attributed to the activities of competitive or antagonistic microorganisms as well as the antibiotic compounds present in the vermicompost.

Table. List of Fungi Isolated from Partially Decomposed Dry Organic Waste for Vermicompost and Vermicompost.

Partially decomposed dry organic waste for vermicompost	Vermicompost
Alternaria citri	Absidia cylindrospora
Aspergillus fumigatus	Aspergillus fumigatus
Aspergillus niger	Aspergillus niger
Aspergillus cervinus	Aspergillus clavoto nanicus
Aspergillus terreus	Aspergillus terreus
Aspergillus sydowii	Aspergillus sydowii
Aspergillus niveus	Aspergillus nidulans
Aspergillus sclerotiorum	Cladosporium herbarum
Cladosporium cladosporioides	Fusarium oxysporum
Cladosporium herbarum	Fusarium semitactum
Fusarium samucinum	Fusarium nivale
Fusarium dimerum	Mucor circinelloides
Mucor racemosus	Stachbotrys chartarum
Penicillium chrysogenum	Trichoderma viride
Penicillium thomii	
Penicillium citrinum	
Trichoderma viride	
Thermomyces lanuginous	

Vermicomposting: A Livelihood Micro-enterprise for Rural Women

ICRISAT with support from the Asian Development Bank (ADB), Philippines, District Water Management Agency (DWMA), Government of Andhra Pradesh and Tata-ICRISAT-ICAR project in northeastern regions of India was keen to promote the vermiculture technology.

The primary objective of this project was to help women from rural areas to set up micro-enterprises based on vermiculture technology and also to improve crop productivity by increasing soil fertility through ecological methods of farming.

The training programme conducted by ICRISAT for DWACRA (Development of Women and Child in Rural Area) group of women and other women self-help groups (SHGs) covered technical aspects of multiplying earthworms, managing and collection of organic wastes, application of vermicompost for various crops, accounting and marketing.

At the same time a noxious weed, *Parthenium hysterophorus*, was found abundantly in the fields as well as on field bunds, which inhibited crop growth and caused environmental pollution. Hence, the women have come forward to utilize this weed as raw material for vermicomposting, which is a safe weed disposal mechanism and an opportunity to convert into valuable compost.

Adarsha Watershed, Kothapally

Ms Lakshmamma and four other women have set up a vermicomposting enterprise in a common place under one roof. Having begun with a population of 2,000 earthworms of three epigeic species, they regularly harvest around 400 kg of vermicompost every month collectively.

Their work in making vermicompost is shared collectively and the unique marketing strategy involves meeting potential customers. Sometimes, they even get customers from distant places. They earn a net income of around Rs 500 each month. By becoming an earning member of the family, they are involved in the decision-making process in the family. This has also raised their status in the society.

APRLP Watershed Village

Ms Padmamma living in Sripuram, one of the thousand non-descript villages of Mahbubnagar district in Andhra Pradesh, leads a routine life and has never dreamt of a different life. She joined the women's SHG at the begining of the Andhra Pradesh Rural Livelihoods

Programme (APRLP) project. Though reluctant during the initial stage, she started taking active part in the weekly meetings and showed interest in the discussions about raising income through small initiatives like adopting the vermicompost scheme.

This scheme was introduced to enhance crop productivity in the fields and enable the farmers to get more per-hectare yield. Ms Padmamma is able to get higher yield from different crops such as maize and vegetables with the application of vermicompost in her own field. She now proudly displays the vermiculture beds to any visitor who comes to meet her.

Tata-ICRISAT-ICAR Project

The farmers of Bundi nucleus watershed in Rajasthan, India have shown lot of interest in vermicomposting. Two farmers have built a multiple compartment system (commercial model) of vermicomposting while many are following the regular vermicomposting. In Guna nucleus watershed in Madhya Pradesh, nearly 35 farmers from all the three microwatersheds are practicing vermicomposting.

Most of them are producing vermicompost on a large scale and are applying to their own fields for vegetable crops and getting higher yields with low-cost technology. A few farmers have already started selling their extra produce of vermicompost at the nearby market at the rate of Rs 5–7 per kg.

The production of degradable organic waste and its safe disposal becomes the current global problem. Meanwhile the rejuvenation of degraded soils by protecting topsoil and sustainability of productive soils is a major concern at the international level. Provision of a sustainable environment in the soil by amending with good quality organic soil additives enhances the water holding capacity and nutrient supplying capacity of soil and also the development of resistance in plants to pests and diseases.

By reducing the time of humification process and by evolving the methods to minimize the loss of nutrients during the course of decomposition, the fantasy becomes fact. Earthworms can serve as tools to facilitate these functions. They serve as "nature's plowman" and form nature's gift to produce good humus, which is the most precious material to fulfill the nutritional needs of crops. The utilization of vermicompost results in several benefits to farmers, industries, environment and overall national economy.

To farmers:

- Less reliance on purchased inputs of nutrients leading to lower cost of production

- Increased soil productivity through improved soil quality
- Better quantity and quality of crops
- For landless people provides additional source of income generation

To industries:

- Cost-effective pollution abatement technology

To environment:

- Wastes create no pollution, as they become valuable raw materials for enhancing soil fertility

To national economy:

- Boost to rural economy
- Savings in purchased inputs
- Less wasteland formation

Chapter 6

Restoration of Degraded Ecosystems in Asia

The restoration and rehabilitation of degraded ecosystems is an important component of nature conservation throughout the world in 1990's. The long-term goal (in about 40 years) is to restore over 30% of degraded lands in each biogeographical region by means of modern science and technology.

The short-term goal (in about 10 years) is to create appropriate planning, in order to restore sustainable utilization of representative degraded habitats of main ecosystem types, particularly those which have been seriously affected by aridity, desertfication, deforestation and soil erosion. This paper mainly reviews the present situation and developmental tendency of land use after deforestation in tropical and subtropical China with a view to providing some reference for reasonable planning of land use in the future.

FORMATION AND UTILIZATION OF DEGRADED HABITATS

GEOGRAPHICAL SCOPE AND NATURAL FEATURES

The tropics of China are located at the northern margin of the Asian tropics. They include a vast extent from the southern Wanshan islands in Guangdong at 114°E in the east, to the Yadong, Nielamu of southeast Xizang at 85°E in the west. The northern boundary of the tropics in China is rather low because of the influence of the cold wave that occurs in winter. Their northern boundary is situated at 22°-25°N in central Guangdong. In southwestern Guangxi the tropics are found as far north as 24°N, owing to the following three reasons:

- The southeast monsoon proceeds up along the Youjiang river;
- The northern mountains hinder the cold wave from proceeding farther south;
- Influence by the limestone stratum.

The tropics in southwest Yunnan go as far as 25°N due to the effect of the Bengalese warm air mass. In southern Xizang the tropical zone even goes as far as 28°-29°N; this is the most northern tropical region in China.

The geology and geomorphology throughout China's tropics change in a complicated pattern. There are alluvial plains, coral islands, valleys, hills, mountains, plateaus, limestone peak forests and shrublands, etc. The topography ascends gradually from east to west. The eastern part is characterized by hills and mountains along the coast. Here, the relief gently slopes, most of the area is under the elevation of 150m, and in general the hills and mountain rise 500-1000m in elevation.

The highest point is Wuzhishan mountain in Hainan with an elevation of 1879m. Throughout southwestern Guangxi are limestone mountains, mixed with granite, sandstone and shale mountains, with elevations spanning 200-700m. Some of these mountains reach elevations of over 1000m. There are also coral islands, called the South Sea Islands, distributed widely in the South China Sea.

The western part of the tropics incorporates the southern margin of Yunnan plateau and the southeastern slope of the Himalayas. In general, the elevation of river valleys and basins in southern Yunnan are between 300-600m, while the mountains rise to between 1000-1500m in elevation.

Southeastern Xizang is composed of high mountains and deep valleys with great relief and steep topography. Here, river valleys are generally under the elevation of 200m, while nearby summits may reach over 5000-6000m in elevation.

The climate belongs to the tropical monsoon climate with high temperatures, and average temperature is 20o-28°C. The annual average temperature of the coldest month (January) ranges from 12° to 15°C. The absolute lowest temperature is above 5°C — as averaged over many years, but low temperatures under 0°C may occur in some years. Annual precipitation throughout most of the area is between 1000 to 1500 mm.

The flora belongs to a part of the India-Malaysia florist region and contains many ancient plants, some of which represent an important part of the composition of the modern vegetation. The eastern tropics are mainly composed of components of the eastern flora of Malaysia mixed with some Sino-Japanese and Chinese tropical endemic plant species.

The western tropics are mainly composed of India-Himalayan and Chinese tropical endemic plant species.

The zonal vegetation type is seasonal rain forest, which is replaced by monsoon forest after it is damaged or in dry regions. There are apparent vertical vegetational distributions in mountain areas, and this zonation is also affected by latitude. Vegetation types are rich and colourful, varying dramatically in high mountain and deep valley regions throughout the western tropics.

THE PRESENT SITUATION AND PROBLEMS OF LAND USE

There has been a long history of shifting agriculture with slash and burn cultivation in this area, like that of other tropical regions in the world. In the past, the population was smaller and techniques of cultivation were primative.

Within a large area of forest, such slash and burn cultivation was sustainable — after trees in a small area were cleared and the area then deserted within 2-3 years, the forest could easily have time to regenerate.

Thus, the forests were not seriously threatened, and the biodiversity was rich and varied owing to the occurrence of serial vegetation which also provided abundant food sources for herbivores, the reproduction of which attracted many predators. Actually, there were a lot of south China tigers in Guoxhou and on the Leizhou peninsula of Guangdong in the 1950's.

They were so harmful for people and livestock that the villagers were scared to stop in the mountains and fields in the morning and evening. But, at present, it is difficult to find their trace, because their numbers declined dramatically when a large area of natural forests were burnt for the cultivation of Brazil rubber trees.

Large mammals also depend on shifting cultivation to a great extent. Because most edible foods in mature tropical forests are found high in the canopy, these resources are out of the reach of terrestrial herbivores. Only cleared areas can provide food for cattle and sheep. Thus, shifting cultivation was suited to and played a significant role in the micro-agricultural economy at that time.

However, with the increase in population and concomitant development of science, a large area of slash and burn cultivation will cause a decline in the productive capability of the land, will damage the normal successional course of vegetation, and will eliminate the capacity of the forest to naturally restore itself.

A large area of grassland may replace the forest. In particular, the increased distribution of grassland dominated by *Imperata cylindrica* not only prevents progressive succession of vegetation, but also hinders the growing of cultivated plants. Although the formation of large areas of rice fields may establish a steady agricultural base, the original forest landscape will be thoroughly transformed and natural biodiversity will decline to a montonous condition.

In order to maintain regional ecological balance, it is necessary to keep a basic ratio and rational spatial pattern between primary forests, productive development of agroforestry, and construction of cities, towns, villages and roads.

Because of a the lack of research on this issue, the ecological environment of many places may be seriously harmed and unable to carry on ecological process normally. The results may be serious, such as soil erosion, water source shortage, decline of cycle capacity of nitrogen and nutrition, dramatic change of regional climate, and acceleration of the extinction rate of species.

In turn, these results may negatively affect the productive development of industry and agriculture, and diminish the living level of inhabitants. Obviously, it is extremely urgent to develop regional or watershed planning by the principle of resources conservation combined with sustainable use. Such a programme will enable the construction of a sustainable society.

In the tropics of China, particularly in the Tai villages of Xishuangbanna, courtyard economy is well managed. Here, timber and fuel trees, woody oil and food trees, fruit trees, medicinal plants and vegetables — in short, everything that one expects to find in a dense natural forest — occur in the hills and mountains around the villages. However, artificial timber aforestration is rarely practiced after cutting of natural forests; most of the timber trees come from wild ones of natural forests.

Owing to the rapid development of tropical economic crops and transformation of the micro-agricultural economy, the natural and court managed model mentioned above has been replaced by a large scale of reclamation.

We have now had several decades to reflect on the advantages and disadvantages of these changes. It seems essential to conduct further research on a model of coexistance between natural resources and modernization, and a sustainable developmental strategy.

THE AFFORESTATION OF LARGE AREAA OF THE BRAZILIAN RUBBER TREE

In the early 1950's, due to urgent needs dictated by political and economic aspects in the world, a large risk was taken: the Brazilian rubber tree (*Hevea brasiliensis*) was introduced and widely cultivated in Hainan and Xishuangbanna, Yunnan. There was no hesitation to sacriface a large area of natural forest in order to get necessary rubber to meet economic and military demands.

From the view point of ecology this introduction produced a series of problems: the plant was at risk because the absolute lowest temperature in winter may go beyond its endurance, and the frequent typhoon, rare and irregular annual precipitation, lower fertility and strong evaporating intensity of soil, etc. were not favorable for its growth.

In a word: cold, wind, and drought calamities were key factors preventing its development over large areas. Hope for success depended on choosing approproate micro-habitats and the practise of intensive cultivation. This predication has been provided by more 40 years of experience. Traditionally, cultivation of the Brazil rubber tree cannot go beyond 17°N.

Nevertheless this barrier has been broken down recently through multiple efforts. Now it is possible to grow the species as far north as 22°30' N, through breeding many new varieties of high quality and with resistance to wind, drought and cold.

The species may now be cultivated and managed with techniques suited to terrace, hill and mountain. The result has been establishment of a new ecosystem of Brazil rubber forest with a shelter forest network and green cover adapted to northern tropical environments.

The largest cultivated area is over 300,000 ha, and China now occupies the fourth position of 28 countries of cultivating Brazil rubber trees in the world. Its yield of dry rubber has reached 100,000 tons, occupying the sixth position in the world. Although serious calamities occur in some places every year, the total yield increases continuously.

Annual mean production per ha and individual tree also has been relatively steady or increasing. There is significant argument about whether it is worthwhile to develop Brazil rubber forests in the economy. Some analyses show that certain areas of natural forest have been damaged to some extent, but that the occurrence of large areaa of artificial forest also have played a great role in safeguarding regional ecological balance and in promoting local economic development and people's living level.

It should be indicated that many Brazil rubber tree plantations were established after deforestation, with many new towns created and new social structures formed. Most of the inhabitants moved to new towns from other places, and their income and living level are higher than that of native people. As a result, a series of ecological, social and economic contradictions have occurred concerning proposals to restrain the development of rubber property.

From the view point of biodiversity conservation and sustainable use, there are many species in the tropics, and many local species are worthy of being developed economically. That is to say, we should not merely pay attention to the Brazil rubber tree in these areas. In order to break away from monotonous landscape composed of a single species or variety, we need to maintain the original primary forests, to concentrate on those resources within tropical regions that may have economic value, and to initiate a diversified economy in line with local conditions.

At present, the development of a market economy has great influences on the development of rubber property in China. In addition, it is difficult to compare products from the distribution centre of the Brazilian rubber tree with those from China, both in price and in quality. Since domestic products cannot compete with foreign ones, economic comparisons are difficult, yet these comparisons unavoidably affect the extent and development of rubber plantations in China. . If government can not effectively regulate this economic endeavor, all that has been achieved will certainly be spoiled.

One hundred million individual Brazilian rubber trees, which have been cultivated over many years, will be damaged. It seems that relevant governmental departments should weigh up the gains and losses to all aspects concerning this introduced tree and make necessary scientific policy.

THE AFFORESTATION OF LARGE AREAS OF EUCALYPTUS

In the process of developing Brazil rubber trees throughout the tropics of China, some locations, such as terraces and hills of north Hainan, south Guangxi and Leizhou peninsula of Guangdong, proved difficult to plant successfully due to damage caused by windstorms, cold and low levels of soil nutrients.

After planting the sugarcane for several years, which resulted in a severe loss of soil fertility, local farmers finally made the decision to cultivate Eucalyptus, which now has engulfed the tropics and subtropics throughout the world. Eucalyptus comes from Australia, it grows rapidly, sprouts intensely and adapts easily to poor soil.

Several Eucalyptus species such as *E. exerta, E. rudis, E. citriodora* and some new varieties, are widely planted. It is true that the timber quality and value of individuals over 20 years old can be high, if they are cultivated with appropriate management measures.

Since most people are anxious to get benefits and to use the wood for paper-industry, continuous cultivation on the same land made soil fertility to become lower and lower. This reduction in soil fertility causes entire forest plantations to become jeopardized. It seems indispensable to carry out sustainable development in this area according to the requirement of biodiversity use and agroforestry.

THE INTRODUCTION OF FOREIGN PINE TREES

There are only few native species of pines that occur in tropical China. *Pinus latteri* and *Pinus fenzeliana* are mainly distributed in the mountainous environment there, although they are very few in number. *Pinus massoniana* has been cultivated in the area, but it grows poorly due to its succeptibility to insect pests and diseases. Several years ago, *Pinus elliotii,* introduced from Gallebi, was introduced. It grows well in dry and poor terraces and hills along the coast.

Its strong wind-resistance attracts people to widely cultivate the species. Some people even intend to replace local *Pinus massoniana* populations with *Pinus elliotti*. Because large areas of pure stands have been planted, serious diseases and problems due to insect pests occurred 10 years after the species was introduced.

The leaves of trees became yellow, and many trees gradually died. In the meantime, the diseases spread very fast. Farmers have had to adapt segregated measures and to control potential outbreaks on their lands. In short, foreign species of pine should not be introduced over large areas without prior long-term observation and experiment.

Rational Distribution of Land Use

The above examples make clear that it is unwise to concentrate on the development of so-called fast growing and high economic trees over wide areas without paying attention to conservation and management of natural forests.

It is absolutely necessary to take advantage of local resources and to choose development of native growing fast, precious and high economic species. As the IUCN-WWF Plant Conservation Plan indicates, it is unwise to concentrate research and development only on a small number of crops and trees such as rice, wheat, Eucalyptus and poplar. We should pay attention to conservation, research and utilization of wild species and their germplasm resources. We should also stress the rational distribution of primary forests, as well as artificial ecosystems, with the aim to form a mosaic, in order to ensure the possibility of sustainable development of China's natural resources.

COASTAL TERRACES AND HILLS REGION IN HAINAN, GUANGDONG AND GUANGXI

The sandy coastal land of this region has formed a shelter forest belt extending several hundred km through the successful introduction of *Casurina equisetifolia,* which in turn has played a great role both in improvement of the regional ecological environment and in the local economy.

It seems beneficial to establish more and better-adapted native trees following the formation of this shelter belt. Such an activity, taking advantage of the pioneer tree and its role in improving the micro-environment, would lead to the construction of a higher quality of shelter belt. (like the forest belt composed of *Vatica hainnanensis* along the coastline of Wanning county in Hainan).

Unfortunately, the litter produced by *Casurina equisetifolia* preclude introductions of native trees. It is not only unable to promote cycling of soil nutrients, but also causes land productivity to degenerate. It is even difficult to grow more *Casurina equisetifolia.* The shelter belt in many locations has been destroyed.

Given this present situation, we ought to increase silvicultural species in line with local conditions, such as with introductions of *Cocos nucifera, Pandanus tectorius, Hibiscus tiliaceus, Acacia confusa, Vatica hainanensis, Micrcos paniculata, Litsea glutinosa, Vitex quinata, Barringtonia asiatica, Hermandia origera, Calophyllum inophyllum* and Ficus spp. , etc.

The terrace and hill environments under the elevation of 500m are mostly occupied by Brazilian rubber forest and Eucalyptus forest, in additon to the rice fields. *Pinus elliotii* forest occurs in some locations. Within the Brazilian rubber forest are found very few species in silviculture, such as *Casurina equisetifolia,* Eucalyptus spp. , and *Acacia confusa.*

In a word, this region features only a dull landscape and few species. At present, an important task of biodiversity conservation and sustainable use is to enrich cultivated areas, to increase the number of species there (and the number of ecosystems composed of them), and especially to enhance the introduction and breeding of native growing species (especially those that occurred there originally and those that grow quickly).

According to long-term observation and experience, the following species are worthy of development in line with local conditions: *Anthocephalus chinensis, Chukrasia tabularis, Melia toosendan, Toona cillata, Toona microcarpa, Choerospondias axillaris, Gmelina hainanensis, Terminalia hainanensis, Vatica hainanensis, Hopea hainanensis, Hopea exalata,* etc. Varied tropical fruit trees should also be encouraged, such as: *Litchi chinensis, Euphoria longan, Mangifera indica, Artocarpus heterophyllus, Clausena lansium, Psidium guajava, Averrhoa carambola, Syzygium jambos,*

Canarium album, Canarium pimela, Dracontomelon duperreanum, Annas comosus and Tamarindus indica, etc.

We should choose some fast-growing species of Leguminosae, such as: Albizzia spp. , Acacia spp. and Leucaena spp. in dry and poor locations. Dwarf wood management should be included in these efforts, especially after the soil has been improved. Research and development projects on fuel wood in southeast Asia are a valuable reference and benchmark for us.

The following species are also merit cultivation: Castanopsis hystrix, Dalbergia odorifera, Homalium hainanense, Madhuca hainanenensis, Michelia mediocris, Alseodaphne hainanensis, Sindora glabra, Manglietia hainanensis and Calamus spp,. ect. in mountains above 500m elevation.

THE LIMESTONE MOUNTAIN REGION IN SOUTHWEST GUANGXI

This region mainly includes the extensive range of the Zuojiang and Youjiang river basins and southeast Yunnan. The strata of this region are composed primarily of limestone mixed with granite, sandstone and shale. The region is higher in the northwest, with elevations between 1000-1300m, and inclines to the southeast where elevations of 100-500m occur. The Daqingshan mountain location in the southern part has some peaks with elevations of 1000m.

This region has high annual accumulated temperatures, although the absolute lowest temerature there may be quite low (due to obstruction and concomitant stagnation of cold air in the southern mountains). The Brazilian rubber tree was planted in this region for nearly 20 years, but the species was difficult to adapt to the local conditions.

When advective radiation frost occurred, individuals growing in the lowlands were seriously damaged and those growing on slopes were dramatically exterminated. As a result, all forests were destroyed completely. Residents in the area now are obliged to plant sisal hemp (*Agave sisalana*). On the contrary, the Brazilian rubber forest can still exist in southeastern Guangxi and western Guangdong. Here, although at a more northern latitude, the cold wave is only weakly affected due to local conditions.

In the limestone region, which is mostly bare mountain with sparse vegetative cover, excellent timber trees such as *Burretiodendron hsienmu* and *Garcinia paucinervis* are cut, and other tree speecies are used as fire wood. The natural forests can be rapidly restored if the hillsides are closed (protected) to facilitate afforestation. In the area, it is a taboo to burn the grass or cultivate waste land, thus after two years soils are washed away and large areas of bare mountain are formed.

Limestone mountains are then not considered as land suitable for silviculture for a long time. In fact, they are abandoned. Practice, however, proves that limestone mountains are suitable for the development of forestry.

In general, the following species can be chosen to plant on limestone mountains under the elevation of 700m: Toona sinensis, Radermachera sinica, Zenia insignis, Deutizanthus tonkinensis, Ailanthus guangxiensis, Alphonsea mollis, Saraca chinensis, Horsfieldia hainanensis, Bischofia javanica, Parashorea chinensis, Burretiodendron hsienmu and Garcinia paucinervis, etc.

The following species can be planted on sites above 700m elevation: Adina racemosa, Wrightia pubescens, Pterospermum truncatolobatum, Cinnamomum saxatile, Lysidice rhodostegia, Toona sinensis and Radermachera sinica, etc. We can cultivate Cephalomappa siensis by use of a dwarf model, and Pennisetum purpureum may be planted in many round depressions, in order to scientifically organize and develop food sources for native cattle and goats.

The following species can be chosen and used on mountains with acid soils under 700m elevation: Anthocephalus chinensis, Melia toosendan, Chukrasis tabularis, Toona ciliata, Hopea chinensis, Vatica guangxiensis, Madhuca subquincuncialis, Eberhardtia aurata and Acrocarpus fraxinifolius, etc.

The following species can also be considered on sites above 700m elevation: Castanopsis hystrix, Erythrophloeum fordii, Mytilaria laosensis, Tsoongiodendron odorum, Michelia macclurei var. sublanea, Michelia hedysosperma, Toona microcarpa, Symingtonia populnea, etc. The above mentioned tropical fruits in the region are also suited to be developed here. IIlicium verum, Cinnamomum cassia are specific species of this region and also worthy to be considered.

THE HILL AND MOUNTAIN REGION IN SOUTHERN YUNNAN

This region mainly includes Xishuangbanna, and southern parts of Honghe and Dehong Autonomous prefectures. The relief within the region rises and falls. The geomorphic structure is mainly composed of plateau mountains with many broad valleys and basins. These are surrounded by circular hills and mountains with elevations from 500-1000m. Some peaks reach elevations of 1300-1500m; the highest peak is above 2000m. The hills and mountains mainly consist of granite, sandstone and shale. Because there are many breaches in southern part, the southwest monsoon can flow into the interior.

When the monsoon meets the towering mountains and the gradually increasing relief, the humid air goes upward and forms abundant rainfall and high temperatures. This effect, in turn, causes the formation at 21oN of a tropical climate. There is a large area of tropical forest, however it was largely destroyed during the development of the Brazilian rubber forest. Shifting cultivation prevails continuously in degraded habitats, and it causes many sites to be very difficult to restore.

There are only a few artificial timber forests, economic forests and fruit plantations. Brazilian rubber forest can be seen everywhere, and its area is rather smaller than Hainan island, occuping the second position in China.

At present, its situation is the same as on Hainan: awaiting a new decision on how to proceed. The cultural landscape is also excessively monotomous for a tropical area. It is an important way to better conserve existing primary tropical forest, to fully utilize the perspective of wildlife resources and to improve the features of degraded habitats.

Therefore, we should develop tropical fruit trees in a big way, establish artificial timber forests, fuel forests and economic forests by use of native fast-growing and precious species, abiding by the principle of ecological agriculture, except setting up agricultural bases of having high and stable production. In the meantime, we should promote the restoration of secondary ecosystems.

Generally speaking, the following species are worthy to be developed at locations under the elevation of 800m: Anthocephalus chinensis, Duabanga grandiflora, Melia toosendan, Gmelina arborea, Chukrassia tabularis var. velutina, Cassia siamea, Dracontomelon macrocarpum, Meusa magassarium, Acrocarpus fraxinifolius, Ailanthus fordii, Pterospermum lanceaefolium, Parashorea chinensis, Vatica fleuryana, Homalium laoticum, Horsfieldia tetratapala, etc.

The following species can be considered on mountains above 800m elevation: Castanopsis hystrix, Schima wallichii, Cinnamomum porrectum, Toona ciliata, Toona microcarpa, Choerospondias axillaris, Michelia hedysosperma, Manglietia microgyne, Manglietia wangii, Camptotheca acuminata, Phoebe macrocarpa and Phoebe puwenensis, etc.

The tropical fruit trees mentioned above should be more widely developed.

FORMATION OF DEGRADED HABITATS IN SUBTROPICAL AND THEIR UTILIZATION

GEOGRAPHICAL EXTENT AND PHYSICAL FEATURES

Owing to the effect of the monsoon from the Pacific and Indian Oceans, the subtropics of China are called humid subtropics of monsoon type, and these are endemic to the region of east Asia.

This subtropical region is different from the subtropics with winter rainfall as found in the Mediterranean region. The humid subtropics of monsoon type in China occupy an area covering about one-forth of the country. Its northern boundary lies from Qinling mountain to the Huaiyhe river at 32°- 34°N.

The southern border is near the Tropic of cancer. The eastern boundary is located at Taiwan and the southeast seashore of the continent and islands around it. The western border includes the east slope of the Xizang plateau down to south of the western edge of Yunnan.

The range of latitude from north to south is 11°-12°. It may be divided into three subzones: the north, the central and the south, because of temperature differences. The extent of longitude strides across 28° from east to west and may be divided into two parts: the east and the west, owing to the difference of rainfall distribution pattern and species composition.

The zoned vegetation is evergreen broadleaf forest in acid soil mountains, and mixed evergreen and deciduous broadleaf forest in limestone mountains. The vertical distribution of vegetation in mountain areas is very apparent, particularly in the area of high mountains and deep valleys in the western part. A series of similar vegetation types appear to the south and north and from low to high elevations.

THE PRESENT SITUATION AND PROBLEMS OF LAND USE

Although the number of species in the humid subtropics in China is less than that found in the tropics, it is superior than that found in the temperate region. There are many endemic species.

For a long time, local people have exploited many economic plant species, which have been widely cultivated, forming an important economic mainstay in the region.

THE ESTABLISHMENT OF ARTIFICIAL TIMBER FORESTS

Managed models of shifting cultivation also exist in the humid subtropics of China. Slash-and-burn cultivation has been commonly used, but people often planted some economic trees when reclaiming and cultivating wasteland (such as *Cunninghamia lanceolata* and *Aleurites fordii).*

When soil fertility declined making it impossible to plant crops, trees were grown which have come to form a woodland. In the meantime, people are also accustomed to afforesting some degraded sites. *Pinus massoniana, Pinus yunnanenssi, Cupressus funebris, Cupressus duclouxii* are in common use.

Recently airplanes have been used to spread the seed of pines, and this method has apparently been successful. It may be said that with the exception of slash-and-burn cultivation in only a few areas, most of this region is increasing in areas of green cover. However, silvicultural quality is not yet high enough. For a period, people were called to establish large areas (over 70 ha) of coniferous forests. The main species were *Pinus massonian, Pinus yunnanensis* and *Cunninghamia lanceolata*.

Opinions have even been expressed that natural forests should be cut to make room for the planting of China fir (*Cunninghamia lanceolata*). In the meantime, when China fir was subject to afforestation, emphasis was often given to reclaimation and cultivation of wasteland, cherishing illusions of becoming useful timber within several years. This species was cultivated from the foot of mountain to the top. Any shrubs and herbs were cleared from the land surface. It seems that efforts were made to eliminate all competitors of the pine.

In fact, many invisible insects and little animals had to switch to catch food from cultivated trees, due to lack of food sources from native plants. As a result, serious diseases and insect pests appeared in China fir forests. So people should be warned to pay attention to this lesson.

They should exploit native fast growing broadleaf species and use them instead of the China fir. Large areas of coniferous forests should not be encouraged. It should be considered as a great achievement that the relic endangered species — *Metasequoia glyptostroboides* has been used as a silvicultural tree species, and it is widely cultivated in moist habitats on plains and banks.

This region also contains the bamboo, *Phyllastachys pubescen.* Large areas of bamboo forest can be found in many places. It is closely related to people's living.

It may be said that it is a plant for making money by local people. Unfortunately, recent extensive cutting has caused a tendency for these forests to decline. It is urgently needed to strengthen the tending of woods — to replant, to enlarge cultivation, and to enable forests to obtain rapid restoration and development.

THE DEVELOPMENT OF ECONOMIC TREES

This region is used to develop diverse economic trees, which mostly are endemic species of China, such as Camellia oleosa, Camellia sinensis, Aleurites fordii, Sapium sebiferum, Rhus verniciflua, Eucommia ulmoides, Cinnamomum camphora, Castanea mollissma, Juglands regia, Magnolia officinalis, Morus alba, Ginkgo biloba, Traachycarpus excelsa, etc. Their cultivation plays a large role in the sustainable use of degraded habitats, and becomes an important part of the development of local economies and enhancement of people's living level.

But, these species are roughly managed because people are more busy with crop cultivation. The decline of *Camellia oleosa* forests is mainly because its cover layer is often uprooted to make compost for crop demand and its cycle of nutrients is destroyed. The problem exists, however, that such development suffers from bad quality, difficulty of transportation, bad sale and low prices, etc.

Therefore, there are ups and downs in the economic uses of such resources, and these change periodically with strong or weak management of the people. In order to solve this problem, at first we need to establish the idea of regional comprehensive agricultural systems and equally to think of developing crop production, forestry, animal husbandry, fisheries, handicrafts, etc. , so that we are able to promote vigorous economic development.

MANAGEMENT OF SECONDARY DECIDUOUS BROADLEAF FORESTS

After zonal vegetation of this region is damaged, a deciduous broadleaf forest stage always appears among the successional stages in the natural restoration process. This successional stage may include *Alniphyllum fortunei* forest, *Styrax japonica* forest, *Liquidambar formosana* forest, *Populus adenopoda* forest, *Betula luminifera* forests, *Carpinus fargesii* forest, *Quercus variabilis* forest, *Quercus fabri* forest, *Quercus acutissima* forest and *Alnus nepalensis* forest, etc.

All these trees have certain economic values. The stability of these forests can be maintained through active forest management if it is determined that there are demonstrable needs from the sustainable use of their products.

If it is determined that there is the need to restore these forests to evergreen broadleaf forests, appropriate measures may also be taken. The deciduous broad leaf forests should not be cut to make room for cultivated crops or planting of China fir. Such replacement can be viewed as stupid, because losses outweigh the potential gains.

ENRICHING BIODIVERSITY IN CULTIVATED AREAS

This region has rich economic plant species, and it is essential that we develop a proper perspective on their use. More species should be chosen for cultivation through experimental study.

And a rational arrangement and distribution scheme for the incorporation of many economical plants in cultivated areas should be put into force to avoid the pattern of dull landscapes promulgated by the cultivation of a single species or variety.

SOUTHERN SUBTROPICS

This region belongs to a transitional area between the subtropics and tropics. It includes an extensive range of hills and mountains in central and southern Taiwan, coastal hills in Fujian and Guangdong, the Zhujiang river delta, hills and mountains in Guangdong and Guangxi, limestone mountains in Yunnan, Guizhou, and Guangxi, middle-elevation mountains and narrow valleys in central and southern Yunnan.

There is a large environmental difference from the east to west, but there are also many common species, which can provide increased development in degraded habitats. In general, the following species are worthy to be considered in hills and mountains composed of acid red and yellow soil: *Toona ciliata, Toona microcarpa, Choerospondias axillaris, Schima wallichii, Michelia macelurea* var. *sublanea, Tsoongiodendron odorum, Symingtonia populnea, Castanopsis hystrix, Erythrophloeum fordii, Mytilaria laosensis, Illicium verum* and *Cinnamomum cassia,* etc.

The following species are suited to be developed in limestone mountain habitats: *Lysidice rhodostegia, Acrocarpus fraxinifolius, Toona sinensis, Zenia insignis, Kalopanax pictus, and Pterospermum heterophyllum,* etc.

CENTRAL SUBTROPICS

This region is typical of the subtropical region, including hills and mountains to the south of Changjiang river, the Sichuan basin, mountains in Guizhou, Hubei and Hunan, Nanling mountains, Miaoling mountains, basins and valleys in the plateau of central Yunnan, high mountains and deep valleys in western Sichuan, and northern and western Yunnan.

This region has complicated environmental conditions. It is one of the mega-biodiversity areas in China and contains many endemic and relic species.

Different places have their own representative species. The

following species, which have wide adaptability, may be provided to be chosen and developed: Liriodendron chinese, Manglietia fordiana, Michelia maudiae, Idesia polycarpa, Bretschneidera sinensis, Pseueosassafras tzumu, Phoebe bournei, Tapiscia sinensis, Camptotheca acuminata, Fagus longipetiolata, Fagus lucida, Castanopsis carlesii, Castanopsis fargesii, Schima argentea, Taiwania cryptomerioides, etc.

The following species are suited to be cultivated in limestone mountain habitats: Keteleeria calcarea, Kalopanax pictus, Broussonetia papyrifera, Cyclobalanopsis glauca, Gleditsia sinensis, Cladrastis platycarpa, Pteroceltis tatarinowii, etc.

NORTHERN SUBTROPICS

This region is a transitional area between the subtropics and warm temperate zone. It includes plains in the lower reaches of the Changjiang river, hills to the north of the Changjiang river, Tongbaishan mountains, Dabeishan mountains, Qinling and Dabashan mountains, etc.

Different places have large environmental differences. Acid yellow brown soil is characteristic of some habitats, and here species such as Kalopanax pictus, Liriodendron chinese, Tapiscia sinensis, Idesia polycarpa, Corylus chinensis, Fagus bashanica, Cercidiphyllum japonicum and Eucommia ulmioides, etc. , occur.

ESTABLISHING BIODIVERSITY CONSERVATION REGIONS AND FIXING RESTORATION OF DEGRADED ECOSYSTEMS IN DIFFERENT REGIONS

When foreign species are introduced to the tropics of China, the result is often a dull landscape, poor species and degraded environment.

Although some attention has been given to the development of some native species in subtropical China, there still exists the need to avoid the domination of single species of introduced flora. Such monocultures cause outbreaks of disease and insect pests, bring about serious water and soil erosion, and lead to a decline of land fertility and loss of natural biodiversity.

It is necessary to give attention to the development of native fast growing broad-leaved trees, to advocate establishment of block or belt mixed coniferous and deciduous forests, to enhance capacity for improvement of soils, to heighten regional environmental quality, to prevent and kill off vectors responsible for diseases and insect pests, and to adopt a requirement of regional economic development.

THE BASIC IMPLICATION OF BIODIVERSITY CONSERVATION REGIONS

Biodiversity conservation regions have been advanced in the process of drawing up the Convention on Biological Diversity. The aim is to carry out the principle to closely combine resource conservation with sustainable use, to realise synchronization of biodiversity conservation and economic construction, to exploit gradually biological resources having economic value and developmental perspective, and to promote economic development. Biosphere reserves and the potential to cluster such reserves put emphasis on the multifunctional roles that may be played by reserves.

These roles include close integration of objectives with those of local people, thus helping them to carry on land planning with rational utilization. Agenda 21 of China points out that 20 biodiversity conservation regions based on representative reserves should be established by the year 2000. All national protected areas and important biogeographical provinces should also choose appropriate sites to set up biodiversity conservation regions.

CRITERIA FOR CHOOSING LOCATIONS OF BIODIVERSITY CONSERVATION REGIONS

A biodiversity conservation region is an examplary site, not all locations are suited for this designation. Such sites need to be chosen appropriately, and the selection should serve as an example for the selection of future sites. Other regions may then absorb experiences and carry on popularization of this initiative in line with local conditions. The main criteria for choosing such a site should be as follows:

The habitat conditions are more complicated and the biodiversity is more varied (including different blocks composed of primary, secondary and artificial ecosystems) than comparative areas. The area should not be smaller than 100,000 ha, and it should be representative of a particular environment.

The main requirement should be for the area to contain key species and indicator species. The area should enable natural ecological process and species evolution to occur, thus ensuring that biodiversity conservation and sustainable use can take place in close combination and at the same time.

The local government must include the formation of the biodiversity conservation region in their projections of economic construction or social developmental planning, and must appropriate this activity with a certain budget. We should establish a specifically managed institution to plan out rational use of such lànds and to define the appropriate scale of investment and other practical details.

Local people must support the comprehensive developmental principle of combining biodiversity conservation with sustainable use and be willing to put their own production into the overall planning of this region.

In a word, the success and failure of a biodiversity conservation region will depend on the common recognition and coordination of local governments, people, and responsible departments of resource management. Relevant manpower and material resources must be organized to sum up experiments in practical processes and to make strenuous efforts for realizing predetermined goals. These objectives must be made continuously, and with constant feedback between all parties.

THE ESTABLISHMENT OF A NETWORK OF BIODIVERSITY CONSERVATION REGIONS TROPICAL AND SUBTROPICAL CHINA

According to the above mentioned criteria, and based on urgent demand, the probable sites of establishment may be chosen as follows.

The Mountains of Central and Southern Hainan

The tropical forests of this region in Hainan are well preserved and contain rich biodiversity. There is a large area of mountains, which can supply development. Some reserves have been established, but only very small parts are actually conserved.

The biodiversity conservation region may be set up based on the Bawangling reserve in Changjiang county, Jianfengling reserve in Yuedong county, Diaoluoshan reserve in Lingshui county, or Wuzhishan reserve in Qiongzhong county.

The Limestone Mountains in Southwestern Guangxi

Large areas of limestone mountain are seldom seen in tropical areas. The nature of flora in this area of limestone mountain belongs to a special part of the India-Malaysia floristic region — called the florist county of borderlands in China and Vietnam.

Here, certain areas of limestone seasonal rain forest are better preserved and with a richer biodiversity than found elsewhere. Many species have a special nature and certain developmental potential. The biodiversity conservation region may be defined based on Longgang reserve in Longzhou county and Longrui reserve in Ningming county.

The Xishuangbanna Region in Yunnan

A large area of primary seasonal rain forest which well preserved and contains rich biodiversity occurs in the Xishuangbanna region. Many wildlife species here have a developmental perspective. Ideally, the biodiversity conservation region should be established around the Mengla reserve, Menglun reserve in Mengla county or Mengyang reserve in Jinghong county.

The High Mountain and Deep Valley Region in Western Sichuan

This region belongs to the mountainous northern end of the Hengduan mountain range. Primary humid subtropical forests are well preserved and of a large area with apparent vegetational vertical distribution here. This is an important region in which biodiversity is concentrated in eastern subtropical China. In the meantime, it is also a famous site for abundant endemic species of China.

Giant pandas are common here. The biodiversity conservation region may be established based on Wolong reserve, in which a special region with governmental functions has been established. Actually, the reserve is also a leading organization for biodiversity conservation regions. It is easy to establish protocols for relevant work if guiding policy is well-defined.

The High Mountain and Deep Valley Region in Western Yunnan

This region belongs to the mountains along the southern margin of the Hengduan mountain range. Primary humid subtropical forests are well preserved in large areas and with apparent representative vertical distribution of vegetation.

This is also an important region, in which biodiversity is concentrated in the western subtropics of China with large amounts of Chinese endemic species. The biodiversity conservation region may be defined on the basis of Gaoligongshan or Nujiang reserves. The region will enable close promotion of economic development of different countries along the Mekong river.

Limestone Mountains in Southern Guizhou

A large area of limestone forest characteristic of the humid subtropics is well preserved in Maolan, Libo county, according to existing information. Here, rich and special elements of plant biodiversity exist, as well as available diverse wildlife resources. Visitors from domestic and foreign countries are attracted to this area by its beautiful Karst scenery.

The biodiversity conservation region may be defined by the Maolan reserve. It will potentially play a great role in the restoration of secondary ecosystems in extensive limestone mountain habitats and in the improvement of local people's living level.

THE SHENNONGJIA REGION IN HUBEI

Shennongjia mountain maintains a certain area of primary subtropical forests in central China with apparent vegetational vertical distribution and rich biodiversity. The biodiversity conservation region here may be established based on the Shennongjia reserve.

Like in the case of the Wolong reserve, it has also set up a government special area and is a leading institution with a definite policy toward establishment and management of biodiversity conservation regions. Its policies carry forward those expressed in the Convention on Biological Diversity.

Nanling Mountain

The Nanling mountain watershed of Changjiang and Zhujiang rivers extends across the contiguous areas of Guangdong, Guangxi, Hunan and Jiangxi. Its flora blends together elements of central, eastern and southern China and contains a rich biodiversity. There are many species that serve as substitutes to those in west Yunnan, located in the western subtropics. The area is rich in endemic and relic species of wildlife resources.

The biodiversity conservation region may be established based on the Huaping reserve in Longsheng county, Miaoershan reserve in Xinan county, Guangxi or two Mangshan reserves by the border of Hunan and Guangdong. It will promote the development of the local economy and enhance the living level of local people.

MOUNTAINS IN ZHEJIANG AND FUJIAN

The subtropical forests of this region are better preserved than any in eastern China, and they contain many endemic species and rich biological resources. The Wuyishan reserve in Fujian has carried on management for several years according to requirements for biosphere reserves, accumulating abundant experience.

This experience can be utilized by other reserves, making it a model for the establishment of biodiversity conservation regions. Baishanzu-Fengyang reserve in Zhejiang has also similar conditions.

Dabeishan Mountain in Western Anhui

The forests of this region are the best preserved of those anywhere in the northern subtropics. They contain elements of rich biodiversity. Many endemic species exist in this area, which acts as a zone of convergence between the flora of southern and northern of China. Here the biodiversity conservation region can be established based on the Yaoluoping or Jinzhai reserves in Anhui. It may be important to promote the development of comprehensive agriculture around the reserves.

It may be predetermined that the establishment and gradual improvement of a network of biodiversity conservation regions can lead to exploitation of biological resources having developmental perspective. It will play great role for the promotion of regional development of industry and agriculture, and march on toward a fine future for the coexistence of nature and modernization in the 21st century.

Chapter 7

Environmental Impacts of Coal Mining

Coal mining is the extraction or removal of coal from the earth by mining. When coal is used for fuel in power generation it is referred to as steaming or thermal coal. Coal that is used to create coke for steel manufacturing is referred to as coking or metallurgical coal.

Coal is also an important source of Methanol which resides in binding resin, plywood, and acetic acid, plastic bottles. Coal is the means of producing more than half of the United States of America's electricity. In the United States, United Kingdom, and South Africa, a coal mine and its accompanying structures are collectively known as a colliery. In Australia, 'colliery' usually only refers to an underground coal mine.

METHODS OF EXTRACTION

The most economical method of coal extraction from coal seams depends on the depth and quality of the seams, and also the geology and environmental factors of the area being mined. Coal mining processes are generally differentiated by whether they operate on the surface or underground. Many coals extracted from both surface and underground mines require washing in a coal preparation plant.

Coal is mined only where mining is technically feasible and economically justifiable. Technical and economic feasibility are evaluated on several factors: regional geologic conditions; overburden characteristics; coal seam continuity, thickness, structure, quality, and depth; strength of materials above and below the seam for roof and floor conditions; topography (especially altitude and slope); climate; land ownership as it affects the availability of land for mining and access; surface drainage patterns; ground water conditions; availability of labour and materials; coal purchaser requirements in terms of tonnage, quality, and destination; and capital investment requirements.

Surface mining and deep underground mining are the two basic methods of mining. The choice of mining method depends primarily on depth of burial and thickness of the coal seam. Seams relatively close to the surface, at depths less than approximately 180 feet (55 m), are usually surface mined.

Coals that occur at depths of 180 to 300 feet (91 m) are usually deep mined but, in some cases, surface mining techniques can be used. For example, some western U. S. coals that occur at depths in excess of 200 feet (61 m) are mined by open pit methods, due to thickness of the seam (60-90 feet). Coals occurring below 300 feet (91 m) are usually deep mined.

MODERN SURFACE MINING

When coal seams are near the surface, it may be economical to extract the coal using open cut (also referred to as open cast, open pit, or strip) mining methods. Open cast coal mining recovers a greater proportion of the coal deposit than underground methods, as more of the coal seams in the strata may be exploited.

Large Open Cast mines can cover an area of many square kilometers and use very large pieces of equipment. This equipment can include the following: Draglines which operate by removing the overburden, power shovels, large trucks in which transport overburden and coal, bucket wheel excavators, and conveyors.

In this mining method, explosives are first use in order to break through the surface of the mining area. The coal is then removed by draglines or by shovel and truck. Once the coal seam is exposed, it is drilled, fractured and thoroughly mined in strips. The coal is then loaded on to large trucks or conveyors for transport to either the coal preparation plant or direct to where it will be used

Most open cast mines in the United States extract bituminous coal. In Australia and South Africa open cast mining is used for both thermal and metallurgical coals. In South Wales open casting for steam coal and anthracite is practiced. Surface mining accounts for around 80% of production in Australia, while in the USA it is used for about 67% of production. Globally, about 40% of coal production involves surface mining.

AREA MINING

Strip mining exposes the coal by removing the overburden (the earth above the coal seam(s)) in long cuts or strips. The spoil from the first strip is deposited in an area outside the planned mining area. Spoil from subsequent cuts is deposited as fill in the previous cut after coal has been removed.

Usually, the process is to drill the strip of overburden next to the previously mined strip. The drill holes are filled with explosives and blasted. The overburden is then removed using large earthmoving equipment such as draglines, shovel and trucks, excavator and trucks, or bucket-wheels and conveyors.

This overburden is put into the previously mined (and now empty) strip. When all the overburden is removed, the underlying coal seam will be exposed (a 'block' of coal). This block of coal may be drilled and blasted (if hard) or otherwise loaded onto trucks or conveyors for transport to the coal preparation (or wash) plant. Once this strip is empty of coal, the process is repeated with a new strip being created next to it. This method is most suitable for areas with flat terrain.

Equipment to be used depends on geologic conditions. For example, to remove overburden that is loose or unconsolidated, a bucket wheel excavator might be the most productive. The life of some area mines may be more than 50 years.

CONTOUR MINING

The contour mining method consists of removing overburden from the seam in a pattern following the contours along a ridge or around a hillside. This method is most commonly used in areas with rolling to steep terrain. It was once common to deposit the spoil on the downslope side of the bench thus created, but this method of spoil disposal consumed much additional land and created severe landslide and erosion problems.

To alleviate these problems, a variety of methods were devised to use freshly cut overburden to refill mined-out areas. These haul-back or lateral movement methods generally consist of an initial cut with the spoil deposited downslope or at some other site and spoil from the second cut refilling the first. A ridge of undisturbed natural material 15 to 20 feet (6. 1 m) wide is often intentionally left at the outer edge of the mined area. This barrier adds stability to the reclaimed slope by preventing spoil from slumping or sliding downhill.

The limitations on contour strip mining are both economic and technical. When the operation reaches a predetermined stripping ratio (tons of overburden/tons of coal), it is not profitable to continue. Depending on the equipment available, it may not be technically feasible to exceed a certain height of highwall. At this point, it is possible to produce more coal with the augering method in which spiral drills bore tunnels into a highwall laterally from the bench to extract coal without removing the overburden.

MOUNTAINTOP REMOVAL MINING

Mountaintop coal mining is a surface mining practice involving removal of mountaintops to expose coal seams, and disposing of associated mining overburden in adjacent "valley fills. " Valley fills occur in steep terrain where there are limited disposal alternatives. Mountaintop removal combines area and contour strip mining methods.

In areas with rolling or steep terrain with a coal seam occurring near the top of a ridge or hill, the entire top is removed in a series of parallel cuts. Overburden is deposited in nearby valleys and hollows. This method usually leaves ridge and hill tops as flattened plateaus. The process is highly controversial for the drastic changes in topography, the practice of creating *head-of-hollow-fills*, or filling in valleys with mining debris, and for covering streams and disrupting ecosystems.

Spoil is placed at the head of a narrow, steep-sided valley or hollow. In preparation for filling this area, vegetation and soil are removed and a rock drain constructed down the middle of the area to be filled, where a natural drainage course previously existed.

When the fill is completed, this underdrain will form a continuous water runoff system from the upper end of the valley to the lower end of the fill. Typical head-of-hollow fills are graded and terraced to create permanently stable slopes.

ROOM & PILLAR MINING

Room & Pillar mining consists of coal deposits that are mined by cutting a network of rooms into the coal seam. Pillars of coal are left behind in order to keep up the roof. The Pillars can make up to forty per cent of the total coal in the seam. Though, this can be abstracted at a later stage

UNDERGROUND MINING

Most coal seams are too deep underground for opencast mining and require underground mining, which method currently accounts for about 60% of world coal production. In deep mining, the room and pillar or bord and pillar method progresses along the seam, while pillars and timber are left standing to support the mine roof.

Once room and pillar mines have been developed to a stopping point (limited by geology, ventilation, or economics), a supplementary version of room and pillar mining, termed second mining or retreat mining, is commonly started. Miners remove the coal in the pillars, thereby recovering as much coal from the coal seam as possible.

A work area involved in pillar extraction is called a pillar section. Modern pillar sections use remote-controlled equipment, including large hydraulic mobile roof-supports, which can prevent cave-ins until the miners and their equipment have left a work area.

The mobile roof supports are similar to a large dining-room table, but with hydraulic jacks for legs. After the large pillars of coal have been mined away, the mobile roof support's legs shorten and it is withdrawn to a safe area. The mine roof typically collapses once the mobile roof supports leave an area.

There are five principal methods of underground mining:

- Longwall mining accounts for about 50% of underground production. The longwall shearer has a face of 1,000 feet (300 m) or more. It is a sophisticated machine with a rotating drum that moves mechanically back and forth across a wide coal seam. The loosened coal falls on to a pan line that takes the coal to the conveyor belt for removal from the work area. Longwall systems have their own hydraulic roof supports which advance with the machine as mining progresses. As the longwall mining equipment moves forward, overlying rock that is no longer supported by coal is allowed to fall behind the operation in a controlled manner. The supports make possible high levels of production and safety. Sensors detect how much coal remains in the seam while robotic controls enhance efficiency. Longwall systems allow a 60-to-100% coal recovery rate when surrounding geology allows their use. Once the coal is removed, usually 75 per cent of the section, the roof is allowed to collapse in a safe manner.
- Continuous mining utilizes a machine with a large rotating steel drum equipped with tungsten carbide teeth that scrape coal from the seam. Operating in a "room and pillar" (also known as "bord and pillar") system—where the mine is divided into a series of 20-to-30 foot "rooms" or work areas cut into the coalbed—it can mine as much as five tons of coal a minute, more than a non-mechanised miner of the 1920s would produce in an entire day. Continuous miners account for about 45% of underground coal production. Conveyors transport the removed coal from the seam. Remote-controlled continuous miners are used to work in a variety of difficult seams and conditions, and robotic versions controlled by computers are becoming increasingly common.
- Blast mining is an older practice that uses explosives such as dynamite to break up the coal seam, after which the coal is gathered and loaded on to shuttle cars or conveyors for removal to a central loading area. This process consists of a

series of operations that begins with "cutting" the coalbed so it will break easily when blasted with explosives. This type of mining accounts for less than 5% of total underground production in the U. S. today.

- Shortwall mining, a method currently accounting for less than 1% of deep coal production, involves the use of a continuous mining machine with moveable roof supports, similar to longwall. The continuous miner shears coal panels 150-200 feet wide and more than a half-mile long, having regard to factors such as geological strata.
- Retreat mining is a method in which the pillars or coal ribs used to hold up the mine roof are extracted; allowing the mine roof to collape as you retreat. This is one of the most dangerous forms of mining owing to imperfect predictability of when the ceiling will collapse and possibly crush or trap workers in the mine.

PRODUCTION

Coal is mined commercially in over 50 countries. Over 7,036 Mt/yr of hard coal is currently produced, a substantial increase over the past 25 years. In 2006, the world production of brown coal and lignite was slightly over 1,000 Mt, with Germany the world's largest brown coal producer at 194. 4 Mt, and China second at 100. 6 Mt.

Coal production has grown fastest in Asia, while Europe has declined. The top coal mining nations (figures in brackets are 2007 estimate of total coal production in millions of short tons)are:

- China (2,804 Mt)
- USA (1,146 Mt)
- India (529 Mt)
- Australia (428 Mt)
- South Africa (283 Mt)
- Russia (347 Mt)
- Indonesia (180 Mt)
- Poland (162 Mt)
- Kazakhstan (103 Mt)
- Colombia (75 Mt)

Most coal production is used in the country of origin, with around 16% of hard coal production being exported.

Global coal production is expected to reach 7,000 Mt/yr in 2030, with China accounting for most of this increase. Steam coal production is projected to reach around 5,200 Mt/yr; coking coal 620 Mt/yr; and brown coal 1,200 Mt/yr.

Coal reserves are available in almost every country worldwide, with recoverable reserves in around 70 countries. At current production levels, proven coal reserves are estimated to last 147 years. However, production levels are by no means level, and are in fact increasing and some estimates are that peak coal could arrive in many countries such as China and America by around 2030.

MODERN MINING

Technological advancements have made coal mining today more productive than it has ever been. To keep up with technology and to extract coal as efficiently as possible modern mining personnel must be highly skilled and well trained in the use of complex, state-of-the-art instruments and equipment.

Future coal miners have to be highly educated and many jobs require four-year college degrees. Computer knowledge has also become greatly valued within the industry as most of the machines and safety monitors are computerized.

In the United States, the increase in technology has significantly decreased the mining workforce from 335,000 coal miners working at 7,200 mines fifty years ago to 104,824 miners working in fewer than 2,000 mines today. As some might see this as a sign that coal is a declining industry its advances has reported an 83% increase of production from 1970 to 2004.

DANGERS TO MINERS

Historically, coal mining has been a very dangerous activity and the list of historical coal mining disasters is a long one. Open cut hazards are principally mine wall failures and vehicle collisions; underground mining hazards include suffocation, gas poisoning, roof collapse and gas explosions. Most of these risks can be greatly reduced in modern mines, and multiple fatality incidents are now rare in some parts of the developed world.

However, in lesser developed countries and some developed countries, many miners continue to die annually, either through direct accidents in coal mines or through adverse health consequences from working under poor conditions. China, in particular, has the highest number of coal mining related deaths in the world, with official statistic 6,027 deaths in 2004.

To compare, 28 deaths were reported in the U. S. in the same year. Coal production in China is twice that in the U. S. ,while the number of coal miners is around 50 times that of the USA, making deaths in coal mines in China 4 times as common per worker (108 times as common per unit output) as in the USA. When compared to industrial countries such as China, the fatality rate is low in the U. S. However, in 2006, fatal work injuries among miners in the U. S. doubled from the previous year, totaling 47.

These figures can in part be attributed to the Sago Mine disaster. The recent mine accident in Utah's Crandall Canyon Mine, where nine miners were killed and six entombed, speaks to the increase in occupational risks faced by U. S. miners.

Chronic lung diseases, such as pneumoconiosis (black lung) were once common in miners, leading to reduced life expectancy. In some mining countries black lung is still common, with 4000 new cases of black lung every year in the USA (4% of workers annually) and 10 000 new cases every year in China (0. 2% of workers). Rates may be higher than reported in some regions.

*Build-ups of a hazardous gas are known as damps, possibly from the German word "Dampf" which means steam or vapou*r:

- *Black damp*: A mixture of carbon dioxide and nitrogen in a mine can cause suffocation.
- *After damp*: similar to black damp, an after damp consists of carbon dioxide and nitrogen and forms after a mine explosion.
- *Fire damp*: consists of mostly methane, a flammable gas.
- *Stink damp*: so named for the rotten egg smell of the sulfur, a stink damp can explode.
- *White damp*: air containing carbon monoxide which is toxic, even at low concentrations

SAFER TIMES IN MODERN MINING

Improvements in mining methods (e. g. longwall mining), hazardous gas monitoring (such as safety-lamps or more modern electronic gas monitors), gas drainage, and ventilation have reduced many of the risks of rock falls, explosions, and unhealthy air quality. Statistical analyses performed by the U. S. Department of Labor's Mine Safety and Health Administration (MSHA) show that between 1990 and 2004, the industry cut the rate of injuries by more than half and fatalities by two-thirds.

However, according to the Bureau of Labour Statistics, mining remains the second most dangerous occupation in America. · New braces called Atlas Cribs contain a mix of hardwoods and a main lateral element that make these braces stronger than other braces used in the past. The new cribbing system takes up 41 per cent less area than existing ones and may be up to 50 per cent more efficient in terms of airflow.

ENVIRONMENTAL IMPACTS

Coal mining can result in a number of adverse effects on the environment. Surface mining of coal completely eliminates existing vegetation, destroys the genetic soil profile, displaces or destroys wildlife and habitat, degrades air quality, alters current land uses, and to some extent permanently changes the general topography of the area mined, This often results in a scarred landscape with no scenic value, though rehabilitation can mitigate some of these concerns.

Mine tailing dumps produce acid mine drainage which can seep into waterways and aquifers, with consequences on ecological and human health. If underground mine tunnels collapse, this can cause subsidence of land surfaces.

During actual mining operations, methane a known greenhouse gas, may be released into the air. And by the movement, storage, and redistribution of soil, the community of microorganisms and nutrient cycling processes can be disrupted.

EFFECTS OF MINING

- Release of carbon dioxide and methane, both of which are greenhouse gases causing climate change and global warming according to the IPCC. Coal is the largest contributor to the human-made increase of CO_2 in the atmosphere.
- Waste products including uranium, thorium, and other radioactive and heavy metal contaminants
- Acid rain
- Acid mine drainage (AMD)
- Interference with groundwater and water table levels
- Impact of water use on flows of rivers and consequential impact on other land-uses
- Dust nuisance
- Subsidence above tunnels, sometimes damaging infrastructure
- Rendering land unfit for the other uses

Coal mining causes a number of harmful effects. When coal surfaces are exposed, pyrite (iron sulfide), also known as "fool's gold", comes in contact with water and air and forms sulfuric acid. As water drains from the mine, the acid moves into the waterways, and as long as rain falls on the mine tailings the sulfuric acid production continues, whether the mine is still operating or not.

This process is known as acid rock drainage (ARD) or acid mine drainage (AMD). If the coal is strip mined, the entire exposed seam leaches sulfuric acid, leaving the subsoil infertile on the surface and begins to pollute streams by acidifying and killing fish, plants, and aquatic animals which are sensitive to drastic pH shifts. Coal mining produces methane, a potent greenhouse gas. Methane is the naturally occurring product of the decay of organic matter as coal deposits are formed with increasing depths of burial, rising temperatures, and rising pressures over geological time. A portion of the methane produced is absorbed by the coal and later released from the coal seam and surrounding disturbed strata during the mining process. Methane accounts for 9% of greenhouse gas emissions created through human activity.

According to the Intergovernmental Panel on Climate Change, methane has a global warming potential 21 times greater than that of carbon dioxide on a 100 year time line. While burning coal in power plants is most harmful to air quality, due to the emission of dangerous gases, the process of mining can release pockets of hazardous gases. These gases may pose a threat to coal miners as well as a source of air pollution.

This is due to the relaxation of pressure and fracturing of the strata during mining activity, which gives rise to serious safety concerns for the coal miners if not managed properly. The buildup of pressure in the strata can lead to explosions during or after the mining process if prevention methods, such as "methane draining", are not taken.

Wherever it occurs in the world, strip mining severely alters the landscape, which damages the the values of the natural environment in the surrounding land. Strip mining, or surface mining of coal completely eliminates existing vegetation, destroys the genetic soil profile, displaces or destroys wildlife and habitat, degrades air quality, alters current land uses, and to some extent permanently changes the general topography of the area mined. The community of micro organisms and nutrient cycling processes are upset by movement, storage, and redistribution of soil.

Generally, soil disturbance and associated compaction result in conditions conducive to erosion. Soil removal from the area to be surface mined alters or destroys many natural soil characteristics, and may reduce its productivity for agriculture or biodiversity. Soil structure may be disturbed by pulverization or aggregate breakdown.

Removal of vegetative cover and activities associated with construction of haul roads, stockpiling of topsoil, displacement of overburden and hauling of spoil and coal increase the quantity of dust around mining operations. Dust degrades air quality in the immediate area, can have adverse impacts on vegetative life, and may constitute a health and safety hazard for mine workers and nearby residents.

The land surface, often hundreds of acres, is dedicated to mining activities until it can be reshaped and reclaimed. If mining is allowed, resident human populations must be resettled off the mine site, and economic activities such as agriculture or hunting and gathering food or medicinal plants are displaced, at least temporarily. What becomes of the land surface after mining is determined by the manner in which mining is conducted.

Surface mining can adversely impact the hydrology of a region. Deterioration of stream quality can result from acid mine drainage, toxic trace elements, high content of dissolved solids in mine drainage water, and increased sediment loads discharged to streams.

Waste piles and coal storage piles can yield sediment to streams, and leached water from these piles can be acid and contain toxic trace elements. Surface waters may be rendered unfit for agriculture, human consumption, bathing, or other household uses. Controlling these impacts requires careful management of surface water flows into and out of mining operations.

EFFECTS ON WATER

Flood events can cause severe damage to improperly constructed or located coal haul roads, housing, coal crushing and washing plant facilities, waste and coal storage piles, settling basin dams, surface water diversion structures, and the mine itself. Besides the danger to life and property, large amounts of sediment and poor quality water may have detrimental effects many miles downstream from a mine site after a flood.

Ground water supplies may be adversely affected by surface mining. These impacts include drainage of usable water from shallow aquifers; lowering of water levels in adjacent areas and changes in flow directions within aquifers; contamination of usable aquifers below mining operations due to infiltration or percolation of poor quality mine water; and increased infiltration of precipitation on spoil piles.

Where coal or carbonaceous shales are present, increased infiltration may result in increased runoff of poor quality water and erosion from spoil piles; recharge of poor quality water to shallow groundwater aquifers; or poor quality water flow to nearby streams. This may contaminate both ground water and nearby streams for long periods. Lakes formed in abandoned surface mining operations are more likely to be acid if there is coal or carbonaceous shale present in spoil piles, especially if these materials are near the surface and contain pyrites.

EFFECTS ON WILDLIFE

Surface mining of coal causes direct and indirect damage to

wildlife. The impact on wildlife stems primarily from disturbing, removing, and redistributing the land surface.

Some impacts are short-term and confined to the mine site; others may have far reaching, long term effects. The most direct effect on wildlife is destruction or displacement of species in areas of excavation and spoil piling. Mobile wildlife species like game animals, birds, and predators leave these areas. More sedentary animals like invertebrates, many reptiles, burrowing rodents and small mammals may be directly destroyed.

If streams, lakes, ponds or marshes are filled or drained, fish, aquatic invertebrates, and amphibians are destroyed. Food supplies for predators are reduced by destruction of these land and water species. Animal populations displaced or destroyed can eventually be replaced from populations in surrounding ranges, provided the habitat is eventually restored. An exception could be extinction of a resident endangered species.

Many wildlife species are highly dependent on vegetation growing in natural drainages. This vegetation provides essential food, nesting sites and cover for escape from predators. Any activity that destroys this vegetation near ponds, reservoirs, marshes, and wetlands reduces the quality and quantity of habitat essential for waterfowl, shore birds, and many terrestrial species.

The commonly used head of hollow fill method for disposing of excess overburden is of particular significance to wildlife habitat in some locations. Narrow, steep sided, V shaped hollows near ridge tops are frequently inhabited by rare or endangered animal and plant species. Extensive placement of spoil in these narrow valleys eliminates important habitat for a wide variety of species, some of which may be rendered extinct.

Broad and long lasting impacts on wildlife are caused by habitat impairment. The habitat requirements of many animal species do not permit them to adjust to changes created by land disturbance. These changes reduce living space. The degree to which a species or an individual animal tolerates human competition for space varies. Some species tolerate very little disturbance. In instances where a particularly critical habitat is restricted, such as a lake, pond, or primary breeding area, a species could be eliminated.

Large mammals and other animals displaced from their home ranges may be forced to use adjacent areas already stocked to carrying capacity. This overcrowding usually results in degradation of remaining habitat, lowered carrying capacity, reduced reproductive success, increased interspecies and intraspecies competition, and potentially greater losses to wildlife populations than the number of originally displaced animals.

Degradation of aquatic habitats has often been a major impact from surface mining and may be apparent to some degree many miles from a mining site. Sediment contamination of surface water is common with surface mining. Sediment yields may increase 1000 times over their former level as a direct result of strip mining. In some circumstances, especially those involving disturbance of unconsolidated soils, approximately one acre foot of sediment may be produced annually for every 80 acres (320,000 m^2) of disturbed land.

In some situations, surface mining may have beneficial impacts on some wildlife. Where large, continuous tracts of forest, bush land, sagebrush, or grasslands are broken up during mining, increased edges and openings are created. Preferred food and cover plants can be established in these openings to benefit a wide variety of wildlife. Under certain conditions, creation of small lakes in the mined area may also be beneficial.

These lakes and ponds may become important water sources for a variety of wildlife inhabiting adjacent areas. Many lakes formed in mine pits are initially of poor quality as aquatic habitat after mining, due to lack of structure, aquatic vegetation, and food species. They may require habitat enhancement and management to be of significant wildlife value.

LOSS OF TOPSOIL

Removal of soil and rock overburden covering the coal resource, if improperly done, causes burial and loss of top soil, exposes parent material, and creates vast infertile wastelands. Pit and spoil areas are not capable of providing food and cover for most species of wildlife. Without rehabilitation, these areas must go through a weathering period, which may take a few years or many decades, before vegetation is established and they become suitable habitat.

With rehabilitation, impacts on some species are less severe. Humans cannot immediately restore natural biotic communities. We can, however, assist nature through reclamation of land and rehabilitation efforts geared to wildlife needs. Rehabilitation not geared to the needs of wildlife species, or improper management of other land uses after reclamation, can preclude reestablishment of many members of the original fauna.

Surface mining operations and coal transportation facilities are fully dedicated to coal production for the life of a mine. Mining activities incorporating little or no planning to establish postmining land use objectives usually result in reclamation of disturbed lands to a land use condition not equal to the original use.

Existing land uses such as livestock grazing, crop and timber production are temporarily eliminated from the mining area. High value, intensive land use areas like urban and transportation systems are not usually affected by mining operations. If mineral values are sufficient, these improvements may be removed to an adjacent area.

COAL STEAM FIRES

Fires sometimes occur in coal beds underground. When coal beds are exposed, the fire risk is increased. Weathered coal can also increase ground temperatures if it is left on the surface. Almost all fires in solid coal are ignited by surface fires caused by people or lightning. Spontaneous combustion is caused when coal oxidizes and air flow is insufficient to dissipate heat, but this occurs only in stockpiles and waste piles, not in bedded coals underground.

Where coal fires occur, there is attendant air pollution from emission of smoke and noxious fumes into the atmosphere. Coal seam fires may burn underground for decades, threatening destruction of forests, homes, schools, churches, roadways and other valuable infrastructure. Spontaneous combustion is common in coal stockpiles and refuse piles at mine sites.

HISTORIC RESOURCES

Adverse impacts on geological features of human interest may occur in a surface mine area. Geomorphic and geophysical features and outstanding scenic resources may be sacrificed by indiscriminate mining.

Paleontological, cultural, and other historic values may be endangered due to disruptive activities of blasting, ripping, and excavating coal. Stripping of overburden eliminates and destroys all archeological and historic features unless they are removed beforehand.

AESTHETIC EFFECTS

Extraction of coal by surface mining disrupts virtually all aesthetic elements of the landscape, in some cases only temporarily. Alteration of land forms often imposes unfamiliar and discontinuous configurations. New linear patterns appear as material is extracted and waste piles are developed. Different colors and textures are exposed as vegetative cover is removed and overburden dumped to the side. Dust, vibration, and diesel exhaust odors are created, affecting sight, sound, and smell. Some members of local communities may find such impacts disturbing or unpleasant.

SOCIOECONOMIC EFFECTS

Due to intensive mechanization, surface mines may require fewer workers than underground mines with equivalent production. The influence on human populations from surface mining is therefore not generally as significant as with underground mines. In low population areas, however, local populations cannot provide needed labour so there is migration to the area because new jobs are available at a mine.

Unless adequate advance planning is done by government and mine operators, new populations may cause overcrowded schools, hospitals and demands on public services that cannot easily be met. Some social instability may be created in nearby communities by surface coal mining.

Many impacts can be minimized but may not be eliminated entirely by use of best mining practices either voluntarily or to comply with government regulatory programs. Financial incentives to minimize costs of production may minimize use of best mining practices in the absence of effective regulation.

Some temporary destruction of the land surface is an environmental price we pay for utilization of coal resources. The scale of disturbance, its duration, and the quality of reclamation are largely determined by management of the operation during mining.

Mountaintop removal to remove coal is a large-scale negative change to the environment. Tops are removed from mountains or hills to expose thick coal seams underneath, and the soil and rock removed is deposited in nearby valleys, hollows and depressions, resulting in blocked and sometimes contaminated waterways. In some areas of the world, remediation is often delayed for decades.

One of the legacies of coal mining is the low coal content waste forming slag heaps. In addition, all forms of mining are likely to generate areas where coal is stacked and where the coal has significant sulfur content, such coal heaps generate highly acidic, metal-laden drainage when exposed to rainfall.

These liquors can cause severe environmental damage to receiving water-courses. Coal mining releases approximately twenty toxic release chemicals, of which 85% is said to be managed on site.

MINE COLLAPSES

Mine collapses, or mine subsidences, have a potential for major effects aboveground, which are especially devastating in built-up areas.

German underground coal-mining, especially in North Rhine-Westphalia, has damaged thousands of houses, and the coal mining industries have set aside many millions in funding for future subsidence damages as part of their insurance and state subsidy schemes.

In a particularly spectacular case in the German Saar region, another historical coal mining area, a suspected mine collapse in 2008 created an earthquake of force 4. 0 on the Richter magnitude scale, causing some limited damage to houses. Previous smaller earthquakes had been increasingly common. Coal mining was temporarily suspended in the area.

BURNING

Combustion of coal, like any other fossil fuel, occurs due to an exothermic reaction between the components of the fuel source and the components of the air surrounding it. Coal is made primarily of carbon, but also contains sulfur, oxygen and hydrogen. The reaction between coal and the air surrounding it produces oxides of carbon, usually carbon dioxide (CO_2 - an important greenhouse gas) in a complete combustion, along with oxides of sulfur, mainly sulfur dioxide (SO_2), and various oxides of nitrogen (NO_x).

Because of the hydrogen and nitrogen components of air, hydrides and nitrides of carbon and sulfur are also produced during the combustion of coal in air. These could include hydrogen cyanide (HCN), sulfur nitrate (SNO_3) and many other toxic substances.

Further, acid rain may occur when the sulfur dioxide produced in the combustion of coal, reacts with oxygen to form sulfur trioxide (SO_3), which then reacts with water molecules in the atmosphere to form sulfuric acid. The sulfuric acid (H_2SO_4) is returned to the Earth as acid rain. Flue gas desulfurization scrubbing systems, which use lime to remove the sulfur dioxide can reduce or eliminate the likelihood of acid rain.

However, another form of acid rain is due to the carbon dioxide emissions of a coal plant. When released into the atmosphere, the carbon dioxide molecules react with water molecules, to produce carbonic acid (H_2CO_3). This, in turn, returns to the earth as a corrosive substance. This cannot be prevented as easily as sulfur dioxide emissions.

Coal and coal waste products, including fly ash, bottom ash, and boiler slag, contain many heavy metals, including arsenic, lead, mercury, nickel, vanadium, beryllium, cadmium, barium, chromium, copper, molybdenum, zinc, selenium and radium, which are dangerous if released into the environment.

Coal also contains low levels of uranium, thorium, and other naturally-occurring radioactive isotopes whose release into the environment may lead to radioactive contamination. While these substances are trace impurities, enough coal is burned that significant amounts of these substances are released, resulting in more radioactive waste than nuclear power plants.

MERCURY EMISSIONS

Mercury emissions from coal burning are concentrated as they work their way up the food chain and converted into methylmercury, a toxic compoundthat harms people who consume freshwater fish. In New York State, winds bring mercury from the coal-fired power plants of the Midwest, contaminating the waters of the Catskill Mountains.

The mercury is consumed by worms, who are eaten by fish, and then by birds, including bald eagles. As of 2008, mercury contamination of bald eagles in the Catskills had reached new heights. Ocean fish account for the majority of human exposure to methylmercury; the sources of ocean fish methylmercury are not well understood.

HYDROGEOLOGIC CHARACTERISTICS OF RECLAIMED OPENCAST COAL MINE

This hydrologic information used in conjunction with the overburden geochemical data can be used to improve mine drainage predictive models and methods. Hydrologic study will bring in a better understanding of the spoil/OB backfilled material that is contacted by the groundwater and the physical, spatial and sequential nature of this contact. This paper may be useful for special handling of overburden dumps, carbonaceous waste, acid-forming materials and placement of alkaline materials like fly-ash in the mine void of opencast mines.

An understanding of the hydrology of surface mine fill material is important in predicting mine drainage quality after reclamation/ backfilling. However, it is poorly understood and one of the least analyzed aspects of mine drainage prediction. Groundwater is an integral chemical component as it serves as the contaminant transport medium.

Therefore, prediction of postmining drainage quality requires the inclusion of surface mine spoil/OB dumps and groundwater hydrology component in the process. The nature, degree, and duration of groundwater and spoil/backfill material interactions need to be factored into any comprehensive mine drainage predictive method or model.

Recharging waters, moving through the unsaturated portion of backfill, will have intermittent episodes where discrete areas of the mine fill are briefly contacted, whereas groundwater within the saturated zone will have a considerably more consistent and longer contact time with that portion of the fill material. Groundwater within the zone of water table fluctuation will contact nearly all of that fill zone periodically during brief episodes when water levels rise. Groundwater will, in both the unsaturated and saturated zones, chemically and physically react with the fill material that it contacts.

However, mine drainage quality prediction is often based on the assumption of uniform contact with 100 per cent of the fill material and does not take into consideration that groundwater only contacts a limited fraction of the fill. Also, under different hydrologic conditions, the sections of the mine fill material contacted by groundwater can change.

It is important to determine what portions of the minefill are contacted by the groundwater and what is the nature of this contact. Mine fill excavations and aquifer testing indicate that there are areas within backfills which because of very low permeability, allow very little groundwater flow through them. These relatively "dead" areas contribute little to the groundwater system and to the associated mine drainage quality.

CHARACTERISTICS OF MINE FILL GROUNDWATER FLOW SYSTEMS

In the past, the groundwater flow regime in opencast mine spoil of the coalfields has generally received little attention and study. Most individuals have made the assumption that ground-water flow in mine spoil is a porous media system, similar to flow through unconsolidated alluvium. Recent field work and testing of surface mine spoil indicate that this assumption is not completely valid.

Caruccio et al. noted that groundwater flow in the backfill of a surface mine in central West Virginia was highly channelized and that it was not observed until one of these randomly located channels was intercepted.

Based on their physical observations during excavations in mine spoil, they referred to the groundwater flow regime as pseudokarst, where groundwater flows mainly through large voids and conduits. Pseudokarst hydraulic characteristics are similar to the characteristics observed in some karst (carbonate underlain) terrains, however, the mechanism of channel and void formation differs. These types of systems are more descriptively defined by the term double-porosity. In India it has been found that many mine fill areas are having pseudokarst hydraulic characteristics.

These dual aquifer flow characteristics exist because mine spoil is an extremely heterogeneous and anisotropic material. The heterogeneities are created by the processes of mining and reclamation.

During mining and subsequent reclamation, spoil becomes sorted to some extent. When dumped by a rock truck or dragline and regraded by bulldozers, the larger minefill particles tend to roll toward the base of the spoil ridges into the valley between the ridges, while the mid-sized and smaller fragments tend to stay on the sides and top of the spoil piles.

FACTORS INFLUENCING HYDRAULIC CHARACTERISTICS

Lithologic Controls

Lithology of the spoil can influence the hydraulic conductivity in reclaimed mines. Parent rock (overburden) of surface mine spoil in Indian sub-continent is comprised primarily of sandstone, siltstone, and shale. It has been studied that increasing per centages of sandstone (and decreasing per centages of shales) in mine spoil appear to yield higher median hydraulic conductivity values and a narrower range of values.

However, the median hydraulic conductivity increases observed were not statistically significant. The apparent trend is explained by the hydraulic properties of the different lithologies and by mechanisms of mining and reclamation. Sandstone-rich spoil zones tend to have larger fragments than shale-rich zones.

This is because sandstones tend to be well cemented and are better able to resist breakage and weathering. Shales tend to break into smaller fragments during mining and more readily weather and break down to silt and clay-sized particles, which decreases the hydraulic conductivity. An accumulation of clay and silt toward the base of the spoil is often observed in monitoring wells.

Phelps observed that spoil bulk density generally decreases with depth. This appears to be caused by the creation of a significant volume of interstitial voids when the large spoil fragments, commonly sandstone, roll to the bases of spoil piles.

MINING METHODS AND TOPOGRAPHY

Surface mine spoil is a highly heterogeneous and anisotropic medium threefore groundwater flow paths are difficult to determine. However, some general trends have been identified and a few assumptions can be made concerning groundwater flow through mine spoil.

The method of extraction, blasting parameters and fragmentation play a very important role in the hydrologic behaviour of the mine spoils. Direction of mining and the configuration of the backfill can dramatically impact the direction of groundwater flow. It ha been observed that the surface over these highly transmissive spoil valleys is more susceptible to subsidence from piping of fine grained spoil materials because of the substantial amount of groundwater movement.

The location of haul roads across the backfill also can influence groundwater flow. The mine fill underlying haul roads can become highly compacted (less transmissive) from the traffic of vehicles and heavy equipment.

Topography influences groundwater flow in surface mine spoil, because groundwater flows down the hydraulic gradient and topography directly influences the hydraulic gradient. However, the water table is also influenced by other geologic and hydrologic conditions, such as permeability variations, local structure, and the adjacent unmined areas.

IMPACTS OF SPOILAGE

Shortly after regrading of the spoil, differential settling and piping of the finer material begin in the backfill. These subsequent processes contribute greatly to the heterogeneity of spoil and are facilitated by infiltrating surface waters and the water table re-establishment. The "uplift" pressure provided by the rebounding groundwater table may aid the shifting and repositioning of spoil fragments.

It is observed significant settling within a year after reclamation on Illinois surface mines and noted that considerable settling may continue at least 2 to 3 years after reclamation. Aquifer testing by the author indicates that settling within spoil continues, apparently at a lesser rate, even 12 years after reclamation.

Rehm et al. and Moran et al states that changes in hydraulic conductivity in spoil are directly related to the mechanisms and timing of the postmining water table re-establishment.

Hydraulic Conductivity and Transmissivity

Hydraulic conductivity and transmissivity of surface mine spoil exhibit a very broad range of values from location to location and within a mine site. Hydraulic conductivity values generated from testing of 103 monitoring wells from 15 surface mines in northern West Virginia and western Pennsylvania ranged from 4.2×10^{-9} to 7.6×10^{-2} m/s. Weiss and Razem noted hydraulic conductivity values ranging from 1.13×10^{-6} to 1.89×10^{-5} m/s in a surface mine located in eastern

Ohio. The hydraulic conductivity of surface mine spoil in three watersheds in eastern Ohio ranged from 5. 4 x 10^{-8} to 6. 7 x 10^{-6} m/s.

Mine spoil tends to be approximately 100 times more conductive than undisturbed bedrock. The differences in hydraulic conductivity cause differences in water levels between mined and unmined areas. In some cases, perched aquifers on the unmined rock can also account for these water level differences. Water levels in spoil wells were 50% lower than those measured in wells of similar elevation in adjacent unmined aquifers.

This illustrates that premining water level measurements may not be indicative of postmining levels. Herring noted that in the Illinois Basin, spoil is more transmissive than aquifers in unmined overburden. This was based on the observation that more water was entering the active pits from adjacent spoils than was coming from adjacent unmined overburden. Weiss and Razem likewise observed greater conductivities in spoil compared to premining values at a mined watershed in eastern Ohio.

POROSITY

As with testing for hydraulic conductivity, porosity determination in mine spoil is also difficult. Therefore, the amount of published data concerning porosity is limited and many of the empirically derived values were determined in the laboratory. Spoil porosity is important to mine drainage quality prediction in terms of determining groundwater storage volumes and predicting water level changes during the process of recharge or discharge.

These characteristics impact the nature and scope of groundwater contact with different spoil zones. The porosity values for certain surface mine spoil tend to be significantly greater than for the undisturbed overburden. Effective porosity values for fractured-rock aquifers have been estimated to range from 0. 001 to 0. 1%. It is observed in laboratory that in coal-bearing strata the primary porosity range varies from 0. 8 to 9. 4% with a mean of 3. 9%.

GROUNDWATER VELOCITY

Aquifer testing indicates that the groundwater velocity in surface mine spoil is substantially greater than that of the undisturbed overburden. Average groundwater velocities are affected by recharge rate, effective porosity, hydraulic conductivity, head differential and can vary widely depending on site-specific conditions. Determination of groundwater velocity is important in mine drainage prediction because it relates directly to groundwater contact with the spoil and groundwater storage turnover rates. Hawkins and Aljoe measured a groundwater velocity range in the backfill of a reclaimed surface mine in central West Virginia of 1.2×10^{-5} to 4.9×10^{-5} m/s. Caruccio et al. observed similar velocities, ranging from 1.4×10^{-5} to 1.8×10^{-5} m/s, for another reclaimed mine in central West Virginia. A groundwater velocity of 2.0×10^{-5} m/s was determined for a surface mine in eastern Ohio. Ladwig and Campion observed a groundwater velocity of 6.1×10^{-4} m/s at a surface mine in Pennsylvania. A groundwater velocity range of 2.7×10^{-3} to 4.3×10^{-3} m/s was measured in surface mine spoil in eastern Kentucky.

Most of these velocities are below the groundwater velocities commonly measured for true karst aquifers, underground mines, and accentuated fractured rock aquifers. However, the measured groundwater velocities were similar to velocities in unconsolidated glacial sands and gravels. Groundwater velocity measurements from the literature should be considered as a range of values and should be applied as such. Determination of the actual groundwater velocity for a site requires on-site testing, which is not possible for premining prediction.

GROUNDWATER RECHARGE

Initially, after reclamation, diffuse recharge from the surface is generally well below premining levels because of the destruction of soil structure, soil compaction by mining equipment, and low vegetative growth, which tend to promote surface water runoff rather than infiltration. Wunsch and others noted that, during re-excavation, spoil within a few inches of the surface was dry indicating little infiltration was occurring.

Decreases in recharge may also be facilitated by increases in porosity in the unsaturated zone. Flow-duration curves show that receiving streams after mining have reduced base flows, which indicate that recharge is decreased (29% less than premining levels) and surface runoff is increased. After this initial period, as soil structure and vegetation re-establishes, diffuse recharge from the surface begins to increase.

This may coincide with the observed increases in hydraulic conductivity after 30 months, as previously mentioned.

The slow recovery of the water table during this period may be linked to the decreased recharge shortly after reclamation and the increased effective porosity and permeability of the spoil. Some of the recharge from the surface during this early period occurs through discrete openings or voids exposed at the surface. Surface runoff flowing across the mine surface enters the spoil through these exposed voids and flows rapidly downward via conduits to the saturated zone.

The recharging water has a limited contact period within the unsaturated spoil zone. In some instances, this infiltrating water will reappear a short distance away (e. g. , 100 m) as a high-flowing ephemeral spring, but in most cases the water recharges the spoil aquifer and is more slowly released at perennial discharge points.

Experience indicates that these exposed voids continue to receive significant amounts of recharge long after final reclamation, re-establishment of the soil structure, and successful revegetation. Groenewold and Bailey observed that surface water running into these swallets may enlarge them and cause considerable subsidence from piping of the finer grained materials. However, this has not been observed to be a significant problem in the Appalachian coal fields.

Others contend that mining may improve the recharge potential from undisturbed areas. Herring observed that the overall recharge and surface water runoff to reclaimed surface mines in the Illinois Basin were greatly increased. He attributes the increased recharge to the dramatic increase in permeability of the cast overburden. He observed a four fold increase in recharge from mining one half of a watershed in Indiana.

Those two studies did not factor in the impact of mining on the soil horizon as discussed by Razem. Once the infiltrating water has passed through the soil horizon, it appears that the recharge potential is dramatically increased. In the Appalachian basin, surface mine spoil aquifers receive a substantial amount of lateral in-flow from adjacent areas. Adjacent unmined areas (low walls and highwalls) as well as previously reclaimed areas will contribute groundwater to the newly reclaimed site.

Groundwater modeling by Hawkins and Aljoe and Hawkins indicates that groundwater flowing from adjacent areas may be the main source of recharge to the spoil aquifer and that this type of recharge occurs on a more continuous basis at a more consistent but lower rate than the recharge through the exposed surface voids. The surface voids will only recharge the spoil when runoff is occurring. In contrast, lateral recharge is controlled primarily by the hydraulic properties of the adjacent aquifer and the hydraulic gradient.

Under certain conditions, spoil can be recharged from groundwater flow from the underlying strata. If artesian conditions exist beneath the stratum underlying the coal (seat rock), groundwater can flow under pressure via fractures in the intervening strata or through boreholes drilled through the pit floor and recharge the spoil.

Artesian-induced recharge has been observed in a reclaimed surface mine in southern Tennessee.

R&D in European Countries

Several standards have already been developed and followed in European countries with respect to methods of opencast mine filling. Evaluating the potential impact of opencast coal mining on water quality (Groundwater Regulations 1998) an assessment framework for Scotland prepared on March 2004 for the Scottish Environment Protection Agency (SEPA). The principal topics, which it addresses, are as follows:

- The geochemical processes responsible for pollutant release and attenuation in coal-bearing strata affected by opencast mining and identification of possible sources of certain hazardous substances in opencast drainage.
- Mining hydro-geological controls on pollutant loadings, including: the hydrogeology of unworked coal-bearing strata; the hydro-geological behaviour of opencast backfill; and relationships between opencast sites and surrounding groundwater systems.
- Methods for characterizing the pollutant release potential of opencast backfill, including consideration of "static tests" (acid-base accounting), "dynamic tests" (leaching columns and humidity cells), and the limitations of both types of test. Modeling approaches, which incorporate geochemical kinetics and hydro-geological complexity have also been considered.
- Formal risk assessment techniques, most notably the "source-pathway-receptor" evaluation framework, and its adaptation to opencast systems.
- Pollution prevention and treatment technologies, including: prevention/minimization of pollutant release processes; active treatment technologies; and passive remediation technologies.
- Environmental regulations which have implications for Scottish opencast coal mining, including both long-established legislation and emerging EU initiatives relating to mining wastes.

R&D in United States

In US also several standards has already been developed under American Society for Testing and Materials (ASTM) E 50 committee on Environmental Assessment and Subcommittee E 50. 03 on Environmental Risk Management/Sustainable Development/Pollution

Prevention initiated several standards related to the placement of different mine fill materials as well as coal combustion by-products (CCBs) in 1998.

The common goal of these standards is to provide guidelines for appropriate selection, testing and placement techniques when mine is reclaimed by placing different fill materials in the mine cavity and how it will interact with the post reclamation water table.

Each standard has a specific scope based on the type of mine placement. The standards provide guidance on how this information can aid individuals/groups associated with the proper placement of mine fill materials/spoils in various mine settings so that there are less chance of contamination of ground water and better stabilization. .

The U. S. Department of the Interior Office of Surface Mining (OSM) is charged with the responsibility of ensuring that the national requirements for protecting the environment during coal mining are met and making sure the land is reclaimed after it is mined. Surface Mining Control and Reclamation Act (SMCRA) requires the mine operator to ensure that:

- The ground water level should rise after reclamation of the land and there should be overall improvement in moisture content of the soil.
- All toxic materials are treated, buried, and compacted, or otherwise disposed of, in a manner designed to prevent contamination of ground or surface water.
- The proposed land use does not present any actual or probable threat of water pollution.
- The permit application contains a detailed description of the measures to be taken during mining and reclamation to ensure the protection of the quality and quantity of surface and groundwater systems, both on- and off-site, from adverse effects of the mining and reclamation process.
- The rights of present users of such water are protected. Any disposal of CCBs at mine sites must be in accordance with those standards and with applicable solid waste disposal requirements.

Estimates of the additional costs to opencast developers of implementing this framework have been prepared; these are considered marginal in relation to the overall costs of site exploration and planning. In India although some of the open cast mines are doing simultaneous filling and reclamation as a part of mining process, the hydrological impact studies of the backfill material are not done.

The prior hydrological studies of the mine site are mainly

concentrated on the virgin land only. However, the studies within the unconsolidated reclaimed land are not done anywhere in India. There are hardly any standard methods of filling the open cast voids in a scientific manner.

This paper may be useful to study the hydrologic regime in a surface mine backfilled area reclaimed by overburden dumps and coal-ash. The referred works are mostly from US & European authors. The topics written here should be viewed as a potential area of research in India.

Chapter 8

Tool for Mine Spoil Restoration

VAST areas of land all over the world have been rendered unproductive by human activities. The situation is particularly alarming in tropical areas where forest loss and degradation, as well as degradation of land that earlier supported forest, are proceeding at unprecedented rates. Ecosystem destruction by mining for coal, quarrying for minerals, and other processes to meet demands of industries, is an inevitable part of civilization.

The increasing human need for these resources will certainly accelerate further degradation of natural habitats, as most of the mining areas are on the land which was previously occupied by forests. All these will lead to acceleration of erosion of biological diversity and creation of several other environmental problems.

THE MINE ENVIRONMENT

As a result of mining and coal combustion, significant areas of land are degraded and existing ecosystems are replaced by undesirable waste materials in the form of dumps, tailing dams and ash dams. The mineral extraction process drastically alters the physical and biological nature of a mined area.

Strip-mining, commonly practiced to recover coal reserves, destroys vegetation, causes extensive soil damage and destruction and alters microbial communities. In the process of removing desired mineral material, the original vegetation is inevitably destroyed and soil is lost or buried by waste.

We are usually confronted with a complete absence of soil, in either a pedological or a biological sense, and what is left is just a skeleton full of limiting factors. Strip-mining can cause compaction, changes in soil texture, loss of soil structure and reduced water infiltration. In addition, steep-sided soil piles are prone to erosion.

Establishment of vegetation on abandoned mined lands is hindered by physical factors such as high temperature, low availability of soil moisture, uncertain structure and unstable slopes due to hilly terrain and compactation. Particularly, in arid and semi-arid areas, limited rainfall during the growing season and high surface temperatures often limit plant establishment and growth. Sparse vegetation growth on abandoned mine soils also results from low organic matter; low levels of plant nutrients, particularly P and N, K concentration; and high levels of metals (Al and Mn).

In certain areas, the main factor in preventing vegetation from becoming established is acidity. During strip-mining, strata in the overburden are exposed, that may contain iron disulfide (FeS_2, pyrite) minerals. These minerals, when exposed to air and moisture, oxidize to produce acid and soluble salts.

The final pH is affected by the amount and type of pyrite present and the acidity of the various potential buffering systems. Pyrite oxidation and hydrolysis give rise to large amounts of H^+ ions which, by decomposition and exchange reactions with other spoil minerals, can give rise to high concentration of Al, Mn, Fe, Zn and Cu. Therefore, toxic concentrations of Al, Mn and Fe may occur where pH is low. Unreclaimed or poorly reclaimed lands are often barren, because problematic soil material left on the surface after mining would not support plant growth.

This creates problems, as waste dumps can be unsightly and subject to erosion and leaching if left unvegetated. In such circumstances, extreme degradation can easily spread beyond the disposal site. This necessitates initiation of restoration of the degraded environment.

ECOLOGICAL RESTORATION

Widespread loss of production and conservation values of natural habitats due to mining operations make largescale ecosystem restoration an increasingly urgent task. Ecological restoration is the process of repairing damage caused by humans to the diversity and dynamics of indigenous ecosystems. There has been a tremendous upsurge in restoration as a technique for reversing habitat degradation worldwide.

Ecological restoration implies that we wish to restore organisms and their interactions with one another and with the physical environment. Ecological restoration concentrates on processes such as persistence of species through natural recruitment and survival, functioning food webs, system-wide nutrient conservation via relationships among plants, animals and the detrivore community.

The goal of restoration is usually to develop a long-term sustainable ecosystem native to the area where mining occurred.

Restoration (= rehabilitation) aims to return the degraded system to some form of cover that is protective, productive, aesthetically pleasing, or valuable in the sense of conservation.

Key processes in restoration include identifying and dealing with the processes leading to degradation in the first place, determining realistic goals and measures of success, developing methods for implementing the goals and incorporating them into land management and planning strategies, and monitoring the restoration and assessing its success. Dobson and coworkers feel that ecological restoration will continue to provide important insight into the way that ecological communities are assembled and ecosystems function. There are many approaches to land and vegetation rehabilitation. These depend on severity of damage to the land resource, the goals of rehabilitation and the availability of resources for repairing the damage.

The most intensive rehabilitation is sometimes termed reclamation, because the damage to sites is so severe that soils have to be replaced and landscapes may have to be reshaped. Reclamation is the process by which derelict or highly degraded lands are returned to productivity, and by which some measure of biotic function and productivity is restored.

Long-term mine spoil reclamation requires the establishment of stable nutrient cycles from plant growth and microbial processes. However, the reclamation process often becomes arrested because of severe limitations in the site or the biota. In summary, rehabilitation, reclamation and restoration can be thought of as a continuum of outcomes from the least to the most similar to the predisturbance ecosystem.

RESTORATION OF MINE SITES

The large-scale land disturbances associated with mining operations and related concerns about the environmental effects have triggered an increasing number of rehabilitation programmes which aim for the restoration of natural ecosystems disturbed by mining. Restoration of mine sites often entails amelioration of physical and chemical characteristics of substrate and ensuring the return of vegetation cover.

If specific problems hindering ecosystem redevelopment can be identified, a cure can be designed using or mimicking natural processes. According to Dobson *et al.* , this process of identification and intervention is the essence of ecological restoration.

The most common response to land degradation has been abandonment or reliance on natural succession to restore lost soil fertility, species richness and biomass productivity. However, the process of natural succession on surface-mined soils is slow due to the removal of topsoil, resulting in elimination of soil seed bank and root stocks and due to soil profile disturbances. As many asoryears can elapse before a satisfactory vegetation cover develops on mine waste. Redevelopment of advanced communities may take a millennium or more.

An important goal of ecological rehabilitation is to accelerate natural successional processes so as to increase biological productivity, reduce rates of soil erosion, increase soil fertility and increase biotic control over biogeochemical fluxes within the recovering ecosystems.

Analysis of different natural successions on natural and artificial substrates suggests that one of the important factors limiting the rate of development is the process of immigration of taxa. There are genuine difficulties in appropriate species reaching a particular site, especially if they have heavy seeds, unless they already occur in the immediate vicinity.

Artificial revegetation is often used to facilitate the generally slow natural rehabilitation process. Artificial seeding of grasses and legumes or both has been a commonly used method to stabilize unconsolidated mine tailings and to encourage natural invasion of tree and shrub seedlings. This ultimately improves site fertility and moisture retention capacity. Once the abandoned mine lands have vegetation growing on the surface, the regeneration of these areas for productive use has begun and offsite damages are minimized.

In addition, establishment of the vegetation on an abandoned mine land also improves the aesthetics of the area. Overburden is the geologic material above coal seams and below the developed soil horizons. Buried seeds and rhizomes are normally absent in overburden. This fact makes seed reserves in the topsoil an important resource that, if handled correctly, can be used successfully to recover disturbed areas by natural vegetation. Since most of the soil seed reserves are found in the surface 5 to 10 cm, upper 5–10 cm topsoil is recommended to be removed and replaced on the top of overburden material.

However, the collection, storage and use of topsoil for restoration of mine areas are limited in many parts of the world. Therefore, recent reclamation strategies have centred on creating soil that will support short-term establishment of native plant species and will sustain long-term successional development.

Harrington describes plantation-related activities used in restoration of damaged sites. According to him, the first step should be stabilization of soil surface by contours, debris dams, mulch, etc. Compaction of soil also needs to be reduced by mechanical disruption.

If needed, macroporosity of the soil can be improved by incorporation of wood and shale. Soil toxicity in terms of pH, metals and salts has to be reduced by suitable amendments and plantation of resistant species and cultivars. Vegetation establishment and mulching with organic matter will increase soil organic matter.

Nutrient limitations imposed for the growth and establishment of vegetation can be minimized by fertilization and other soil amendments. One aspect in vegetation plantation is to improve microsite condition which can be done by scalping, microsite preparation or by clump planting. Once suitable vegetation starts growing, herbivory and physical damage can be reduced by controlled access, fencing and trapping.

PLANTATIONS ON MINE SPOIL

Plantation is the oldest technology for the restoration of lands damaged by human activity. A primary objective for achieving satisfactory rehabilitation of a mined landscape is to establish a permanent vegetation cover. There is an increasing evidence that forest plantations can play a key role in harmonizing long-term forest ecosystem rehabilitation or restoration goals with near-term socioeconomic development objectives.

Plantations can play a critical role in restoring productivity, ecosystem stability and biological diversity to degraded areas. Relative to unplanted sites, plantations have a marked catalytic effect on native forest development (succession) on severely degraded sites. Numerous studies have demonstrated that land rehabilitation benefits from plantations because it allows to jump-start succession.

The catalytic effects of plantations are due to changes in understorey microclimatic conditions (increased soil moisture, reduced temperature, etc.), increased vegetational–structural complexity, and development of litter and humus layers that occur during the early years of plantation growth.

The development of a plantation canopy can alter the understorey microclimate and the soil physical and chemical environment to facilitate recruitment, survival and growth of native forest species. Otherwise, native species would only very slowly, if ever, regenerate on degraded sites. Thus, plantations may act as 'foster ecosystems', accelerating development of genetic and biochemical diversity on degraded sites.

Plantations have an important role in protecting the soil surface from erosion and allowing the accumulation of fine particles. They can reverse degradation process by stabilizing soils through development of extensive root systems. Once they are established, plants increase soil organic matter, lower soil bulk density, moderate soil pH and bring mineral nutrients to the surface and accumulate them in available form.

Their root systems allow them to act as scavengers of nutrients not readily available. The plants accumulate these nutrients and re-deposit them on the soil surface in organic matter, from which nutrients are much more readily available by microbial breakdown. This is exhibited in the levels of available phosphorus and potassium in afforested colliery spots.

Most importantly, some species can fix and accumulate nitrogen rapidly in sufficient quantities to provide a nitrogen capital, where none previously existed, more than adequate for normal ecosystem functioning. Once the soil characteristics have been restored, it is not difficult to restore a full suit of plant species to form the required vegetation.

According to Faulconer *et al.* , other advantages are that establishment of desirable tree species capable of maintaining the site will slow or prohibit invasion of less desirable weedy species, will provide economic returns in the long term, will aid in developing wildlife habitat and will promote hydrologic balance in the watershed.

SPECIES SELECTION FOR PLANTATIONS

The choice of plantation species is likely to greatly influence both the rate and the trajectory of rehabilitation processes. The establishment of a permanent cover of vegetation involves not only growing plants, but it necessitates bringing into being a plant community that will maintain itself indefinitely without attention or artificial aid, and support native fauna. Such performance could be achieved by selecting species adapted to grow, spread and reproduce under severe conditions, provided both by the nature of the damp material and the exposed situation on the dump surface.

The presence of certain tree species in a productive system can result in better soil structure and increased soil nutrient availability. Among species that may be considered suitable for a given degraded site, there may be considerable variations in their capacity to stabilize soils, increase soil organic matter and available soil nutrients and facilitate understorey development.

These variables include susceptibility to pests and diseases, patterns of aboveground and root biomass accumulation, nutrient utilization and allocation, nutrient use efficiency, nutrient retranslocation, litter production and fine root turnover, and rates of litter decomposition and the presence of secondary compounds that may inhibit the activity of decomposing organisms.

While most species appear to act as catalysts for ecosystem rehabilitation, broadleaf species seem to give better results than conifers. Of these, fast-growing species that represent lower successional stages should have preference, particularly those known to establish and grow well on degraded sites.

Where phytotoxicity is suspected, it is particularly important to include plant material from populations growing naturally on mine sites and other areas likely to contain similar toxic factors. In addition to their potential effects on soil fertility, species choices must be guided by seed and seedling availability, local uses for the species and economic aspects. Many studies document the positive role of grass cover as a nurse crop.

Helm emphasized that grasses may have both positive and negative effects on restoration of mine lands. They are frequently needed to stabilize soils during the restoration process, but they may compete with woody regeneration. On the positive side, grasses, particularly C_4 ones, can offer superior tolerance to drought, low soil nutrients and other climatic stresses.

Grasses have fibrous roots that can slow erosion and their sod-forming tendencies eventually produce a layer of organic soil. They are useful in restoration of mined land because they stabilize soil, conserve soil moisture, and may compete with weedy species. This initial cover must allow the development of diverse, self-sustaining plant communities.

Trees can potentially improve soils through numerous processes, including maintenance or increase of soil organic matter, biological nitrogen fixation, uptake of nutrients from below the reach of roots of understorey herbaceous vegetation, increase water infiltration and storage, reduce loss of nutrients by erosion and leaching, improve soil physical properties, reduce soil acidity and improve soil biological activity.

Given time, new self-sustaining topsoils are created by trees. However, impact of trees on soil fertility depends on their nutrient-cycling characteristics such as litter chemistry and decomposition. In addition to the nutrient sink function due to mass accumulation, some plantation species exhibit high nutrient use efficiency and may be more effective nutrient sink than the other species.

In temperate environment, slower-growing, broad-leaved native trees are regarded as better for amenity, but less efficient for timber production. The role of exotic or native species in rehabilitation, needs careful consideration, because we may have to use species combinations (native, exotic or combination thereof) that are capable of surviving into new conditions.

Careful selection of species is needed, as newly introduced exotics may also become pests in other situations. Exotic species are believed to negatively impact site conditions, escape into pristine habitats and displace native species.

Therefore, candidate plantation species should be screened for their potential to become problematic weeds in relation to local and regional floristics. As a part of revegetation efforts, selection of desirable species adapted to the local environment has been emphasized. For artificial introduction, use of species that are well adapted to the local environment should be emphasized.

Indigenous species are preferable to exotics because they are most likely to fit into a fully functional ecosystem and to be climatically adapted. According to Harrington, if there are native species available that are suited to the current soil and site conditions, the regeneration methods have been worked out for the desired species, and the resources are available to cover the cost, then clearly the reestablishment of native species rather than exotics should be preferred.

On mine spoils, nitrogen is a major limiting nutrient and regular addition of fertilizer nitrogen may be required to maintain healthy growth and persistence of vegetation. An alternative approach might be to introduce legumes and other nitrogen-fixing species.

Dobson *et al.* emphasized that the use of nitrogen-fixing species requires good knowledge of their biology, both their soil preferences and their interactions with other species. Nitrogen-fixing species can have a dramatic effect on soil fertility through the production of readily decomposable, nutrient-rich litter and turnover of fine roots and nodules.

Mineralization of N-rich litter from these species will allow substantial transfer to companion species and subsequent cycling, thus enabling the development of a self-sustaining ecosystem. Again, here it should be emphasized that leguminous species also differ in their soil enrichment capabilities.

In a study done for restoration of damaged coal mine areas in India, Singh *et al.* reported that compared to native non-leguminous species, native leguminous species show greater improvement in soil fertility parameters.

Also, native legumes are more efficient in bringing out differences in soil properties than exotic legumes in the short term. In summary, a desired species for planting on mine spoils should possess the ability to:

- Grow on poor and dry soils,
- Develop the vegetation cover in a short time and to accumulate biomass rapidly,
- Bind soil for arresting soil erosion and checking nutrient loss,
- To improve the soil organic matter status and soil microbial biomass, thereby enhancing the supply of plant available nutrients.

In addition, if possible, the species should be also of economic importance.

THE RESEARCH NEED

Strategic and applied research is needed to develop and design management options to facilitate restoration of mined habitats. However, these options need to be developed in full recognition of socio-economic realities, developmental priorities and conservation goals.

Rehabilitation of degraded landscape needs research for identification of stress-tolerant plant species having a positive influence on soil fertility and design of management systems for maximizing ecosystem productivity under a wide range of degraded site conditions. In recent years, interest in rehabilitating land disturbed by mining activity has grown from species selection and site preparation to include research on ecological aspects of reclaimed areas, such as role of organic matter in productivity and nutrient relationships.

The questions that are raised against the use of tree plantations in rehabilitation usually centre on the negative effects of monoculture such as low stability, low resource use efficiency and a low level of biodiversity. Studies are needed to establish biodiversity restoration potential of individual plantation species and of combinations thereof. The pure tree stands offer a unique opportunity to evaluate the nutrient-cycling characteristics of several species that could be useful for their future utilization in land rehabilitation.

Lugo feels that plantations are an ideal subject for ecological studies at the ecosystem level. First, their age and species composition are known and controlled from the time the forest is planted. Second, plantations are simple ecosystems from a structural perspective, and thus provide contrast to the inherent structural complexity of native forests.

Finally, understanding the ecological characteristics is also important because high rates of forest destruction, energy shortages and the need to restore degraded lands to forest production have forced an increase in the rate of plantation establishment in the tropics.

SUSTAINABILITY SCIENCE FOR MINE-SPOIL RESTORATION

Generation of mine waste threatens the economic, ecological and social sustainability of our country. This is one of the visible consequences of imperilled nature– society interactions. Various forms of mining – regardless of whether acknowledged or omitted from studies on land use change, ecological degradation and biodiversity loss – have a role in ecological problems in India. Unsustainable mining, in conjunction with other threats to ecosystems, exacerbates the local environmental degradation, livelihood insecurity and global climate change. In an insightful review, Singh and colleagues propose plantations as a tool for mine-spoil restoration.

Plantations are useful for enhancing the productivity of degraded landscape under a variety of circumstances; nevertheless, they are inadequate mine-spoil restoration tools in isolation. I would argue that to restore mine overburden, a broader vision that incorporates holistic science and policy is called for. Indeed, we shall have to draw on sustainability science, to address the imperilled nature–society interactions as well as to construct a selfsustaining functional ecosystem capable of supporting biodiversity, performing ecosystem functioning and providing ecosystem services to society.

Numerous studies note the evidence of climate change and its implication for terrestrial and oceanic ecosystems in India. In addition to amelioration of local environment, a holistic restoration approach can also address the challenges of global climate change by sequestering large amount of atmospheric carbon dioxide and qualify for the carbon credit earning under the Kyoto Protocol. It is suggested that holistic strategy for restoration of mine-spoil must essentially include the following:

- Policy measures and incentive mechanism to store fertile top-soil layer for use in postmining restoration operation,
- Protection to adjacent refugia, remnant vegetation and ancient trees,
- Attracting seed dispersers,
- Rainwater harvesting,
- Assisted soil remediation through addition of pond-bed silt and sediment as well as earthworm,

- Assistance to available persistent rootstock, if any,
- Direct seeding,
- Vegetative cutting,
- Plantations, such as proposed by Singh *et al.*, as a last option if need be, should be resorted to only in sequential combination with strategies suggested here.

Design and enforcement of mining and restoration policy must accept the role of incentives to retrieve and store topsoil before starting the mining operations. Market mechanisms such as recovering the full cost related to:

- Pre-mining vegetation and soil removal,
- Protection to adjacent vegetation and trees as seed source,
- Post-mining restoration of overburden,
- Treatment to adjacent farmlands and streams affected by mining operations,
- Cost of the preventive measures to mine-induced groundwater pollution, are essential measures that will provide robust incentives to mine owners for minimizing the damage to the environment.

Without such a policy in place, the problem will persist. Throughout the mining operations, protection to adjacent vegetation that is likely to serve as seed source in the vicinity is required. Catalysing natural regeneration of native forest species originating from remnant forests and ancient trees in the vicinity is a useful method in combination with direct seeding and increases likelihood of restoration success.

Remnant vegetation also supports a variety of fauna that help in seed dispersal in adjacent areas. Such a primary succession in post-mining landscapes has been found to be useful. Rainwater harvesting is a crucial step for ecological restoration of mine-spoils.

For example, rehabilitation success to revegetate mine-spoils in arid regions in India was achieved using a combination of rainwater harvesting, soil amendments and plant establishment methods using trees, shrubs and grasses. Such an approach is also socially useful to enhance productivity and availability of products to people.

Addition of organic wastes has been found to increase N fertility at a surface coal mine reclamation site, which ultimately stimulated microbial activity and improved the chemical and physical properties of the reclaimed soil. Silt of the traditional rainwater harvesting systems, called ponds and tanks, is the most effective indigenous soil amendment practice in India.

Pond silt is not only mineral-rich, but also a seed bank for a variety of grasses, herbs, shrubs and trees. Restoration of a variety of wastelands, including mine overburden by direct seeding, planting of large vegetative cuttings, and wherever possible, assistance to persistent root-stock has provided effective results in Rajasthan.

Although persistent rootstock recovery will not be a chief strategy as spoil heaps may be too high to overcome, nevertheless, peripheral dump areas have been reclaimed by assistance to root-stock through a combination of rainwater harvesting. Direct seeding of native species has proved a most useful and cost-effective restoration method.

Seed mixture for direct seeding must be carefully selected based on physical and chemical properties of mine-spoil as well as ecological, economic and social criteria. An useful approach is to include framework species across taxa, herbs, shrubs and trees, early and late successional species, as well as a select few keystone species to accelerate the restoration of a functional ecosystem.

Direct sowing also provides a large base for choice of species. Since it is not necessary to raise the seedling in the nursery, a large number of species can be tried out, compared to planting. Production of seedlings in the forest nursery requires large inputs in terms of time and money. This can be avoided by opting for direct sowing.

Direct sowing is also advantageous as it is comparatively easier to maintain the species mix than in a plantation. It can be combined with planting, natural regeneration or both. Direct sowing helps in enhancing biodiversity per unit area. Direct sowing requires a simple technique for rainwater harvesting in combination with minimal soil work.

Thus, ecologically sound multi-tier vegetation can be effectively developed. In conclusion, I would reiterate that we would also need to draw on the cultural resources such as local knowledge and skills to help address the challenge of mine-spoil restoration. A broader vision alone can address the productivity enhancement of wastelands such as mine-spoil, and can contribute to overall approach for ecological, economic and social sustainability in India. A holistic science and policy suggested here is a step in that direction.

PROMOTING SCIENCE IN INDIA

This is with reference to the three letters by Animesh Chaturvedi, Pushp Deep Pandey and Mohan Karuppayil respectively. These letters are thought-provoking, emphasizing the steps to be taken to improve science education in India.

As a first step, it should be made mandatory that only students who have cleared the national level exams like UGC/CSIR, NET, SLET and GATE will be allowed to join the Ph D programme. This filters off those candidates who join Ph D just to engage themselves in some activity. Candidates admitted both at the national institutes and universities should be provided with a good fellowship and accommodation. Psychological test, as suggested by Chaturvedi combined with a strict eligibility as suggested above will definitely present the best candidates for scientific research.

The opportunities and channels for a scientific career shall be made clear to the students at the postgraduation level itself. For the last many years universities are reeling under a financial crunch to carry out scientific research and hence they introduced self-financing courses. Are the universities improving the syllabi, teaching and research with the money they generate? Or is it being used for the general university expenditure?

The salaries of the faculty and the basic expenditure of the departments are borne by UGC. Then, where does this money go? It should be made mandatory that the money generated by introducing selffinancing courses be made available only to that particular department which generates the money, to improve its facilities.

If there is no improvement in the quality of education imparted for the paying students, then there will be no takers in the long run. Let universities be the centres for teaching excellence. Research in the universities must be confined to faculty who can generate their own money through grants.

Those who cannot get grants can at least take up fulltime teaching instead of running away from both teaching and research and blaming the university for non-availability of funds. In most of the universities, faculty are given a contingency grant every year to run a lab, which is generally not sufficient. Hence, orientation courses for young faculty about the possible sources of funds should be conducted periodically.

To avoid all this, I would like to make a suggestion, which may not please everyone – make universities as centres for high-quality teaching and national laboratories as centres for research. Neither of them will do both teaching and research at the same point of time. Universities nowadays recognize private institutions for postgraduation courses in science. This should be discouraged. The quality and basic values should be preserved. A candidate who joins the university after clearing the entrance exam is given the same degree as a candidate who studies in a private institution.

This is not fair. Let universities alone confer the postgraduate degrees and thus preserve their value. We need postgraduates of quality and not in quantity. Another aspect is the lack of appreciation of importance of research in the general public. For them, scientific research is just another job; they do not know what researchers do. Efforts should be made to popularize the fact that research is the basis for all inventions and even necessities, as in medicine. Urgent steps are to be initiated to stimulate the young minds to opt for science as a career, to see a definite role for India in world science.

MINE SPOIL RESTORATION

Rainwater Harvesting

Rajasthan presents evidence for the existence of one of the most advanced works of ancient mining and accompanied deforestation globally. Mining continues to be an important economic activity in Rajasthan. However, economic benefits of mineral extraction also accompany environmental, economic and social costs. Mine waste dumps and mined out areas viewed simply as the legacies of past may appear overwhelming environmental hazards presenting ugly picture of cultural landscape.

However, mine wastes can be transformed into an opportunity for learning, adaptation and productivity enhancement for sustainable livelihoods through ecological restoration. Here we propose a strategy for mine spoil restoration aimed at creating a multifunctional ecosystem in mine waste dumps. We suggest that dredging and sediment removal from traditional tanks and ponds can potentially be used to prepare the substratum over the mine wastes for direct seeding.

It will also create enhanced decentralized water storage capacity for wildlife and people. Our strategy combines the concomitant revival of traditional water harvesting systems, ground water recharge, enhanced biomass production and an adaptation to random recurrence of droughts in Rajasthan.

Rajasthan, the western state of India, faces a severe challenge of random and persistent occurrence of droughts, water crisis, increasing mine-waste dumps, land use change and degradation of natural resources. In addition, climate change is now a real threat to human society and ecosystems. A holistic restoration is expected to mitigate the impact of climate change by sequestering large amount of atmospheric CO_2 and qualify for the carbon credits and thus livelihoods improvement for participating communities under the Kyoto Protocol.

A strategy that combines options to resolving these issues is need of the hour. Ecological restoration and integrated natural resource management can reduce environmental degradation, augment the productivity of degraded ecosystems and potentially minimize long-term livelihood insecurity of people dependent on natural resources. In this article, we briefly describe traditions of rainwater harvesting as an adaptation to climate change, provide a brief account of ancient and contemporary mining, and propose a novel strategy for mine spoil restoration.

Traditional Water Harvesting Systems

Rajasthan with 10% of India's geographical area and 5% of the country's population is also the driest state. The total surface water resources in the state are only about 1% ofthat of country. The average rainfall is ~550 mm, with per capita water availability of about 700 m^3 –low compared to all India average of about 1200 m^3 per person.

Large parts of Rajasthan face recurrent drought forcing governments periodically to initiate famine relief works. Food and water shortages during these droughts impact men, women and children destroying their assets such as livestock and put livelihoods at risk.

Villages in Rajasthan are notable for traditional water harvesting systems. Small tanks and ponds called *johads, talabs, talai* or *paals* have water spread areas of a few acres, and larger minor irrigation tanks often have command areas of 1000 ha or more. Rajasthan is officially reported to have 4600 such minor irrigation tanks with an estimated potential command area of 0. 63 million ha. These tanks are essential elements in the livelihood adaptations of local communities as aquatic and terrestrial components of tanks provide numerous ecological, social and economic benefits to society.

For example, a traditional runoff harvesting system known as *khadin* in western Rajasthan is practiced where rocky catchments and valley plains occur in proximity. Runoff from the catchments is stored in the lower valley floor enclosed by an earthen bund. Standing water, if any, is discharged through the sluice before crop sowing.

Crops adapted to dry environments are cultivated in *khadin* without any irrigation because soils in *khadins* are extremely fertile and retain residual moisture. Indeed, there is clear evidence from around the world that rainwater harvesting promotes household income, reduces poverty and enhances livelihood security of poor people.

However, these ponds and tanks numbering close to 48,000 small structures are now being neglected by people; they even do not figure in official records. As we discuss later, there is an urgent need for desilting and revival of these multifunctional structures constructed by human societies as an adaptation to climate change over the last 5000 years.

Ancient and Contemporary Mining in Rajasthan

Rajasthan also provides evidence for one of the world's most ancient mining and mineral extraction. Archaeological evidence of ancient mining sites in Rajasthan suggests that the zinc, lead, silver and copper ores were worked extensively in ancient India.

People of Rajasthan were smelting zinc by the sophisticated distillation process at Zawar in Rajasthan by the 12th century AD. A tentative but conservative estimate suggests that 0. 25 million tonnes of zinc concentrates were extracted from some 2. 5 million tonnes of ore in the total mined area before modern mining commenced.

Thus, mines in Rajasthan were important to local people as well as neighbouring Indus-Saraswati civilization, and perhaps other contemporary economies globally. Mining continues to be an important economic activity in India. While acknowledging that reliable estimates of mining area and production are not available, TERI notes that India produces as many as 84 minerals comprising 4 fuel, 11 metallic, 49 nonmetallic industrial and 20 minor minerals.

Interestingly, more than 80 per cent of the mineral production comes from open cast mines resulting into large quantities of overburden. The mining leases numbering 9,244 are spread over 21 States and about 13,000 mineral deposits occupying about 0. 7 million hectares which is 0. 21 per cent of the total land mass of the country.

The aggregate value of the mineral production in 1999-2000 was approximately $10 billion. However, these estimates are difficult to reconcile to take into account the artisanal and small-scale mining in India which has seen large increase with respect to area and the production of minerals.

Not surprising then, at least 90 wildlife sanctuaries and national parks in India with unique biodiversity and wildlife are threatened with mining. Precise estimates of extent of damage is not known and all that can be said with any confidence is that during the course of time about 0. 25 million ha of land has been subjected to varying degree of mining. In Rajasthan there are about 64 different kind of minerals contributing to an annual revenue of more than INR300x108 (US$1=INR43. 5).

Estimates of area affected by mining is yet to become available but may be substantial as there are about 9000 mining units for minerals and more than 16070 quarry licenses employing a total of about 0. 5 million persons per day in the sector. All of the lands may not be owned by government, though. However, it would be useful to initiate the restoration strategy in mines and waste dumps that are situated in officially notified forest lands under the ownership of the government.

THE RESTORATION STRATEGY

Mine spoils left to nature may take decades to centuries to develop any vegetation cover. However, carefully planned artificial interventions that mimic natural processes can reduce this time span. An innovative strategy capable of accomplishing restoration goal has at least 6 key elements. It should:

- Create vegetation cover that may develop into a functional ecosystem in due course of time;
- Provide strong livelihoods improvement incentives to local stakeholders in the form of augmented supply of goods and services;
- Provide an effective framework for cooperation among the stakeholders;
- address both local challenges and the connected global issues;
- Include innovative institutional and financial mechanisms to make it self-enforcing and self-sustaining;
- Maintain options and flexibility for adaptation.

In other words, a holistic science and policy that draws on sustainability science to address the imperiled nature-society interactions is necessary to construct a self-sustaining multifunctional ecosystem capable of supporting biodiversity, performing ecosystem functioning and providing ecosystem services to strengthen livelihoods. A good strategy for mine spoil should also be in coherence with the prevailing policies and actions on water, mineral, wildlife and forest.

It should also essentially provide benefits in terms of climate change mitigation and adaptation. In coherence with these key elements, we propose a strategy for mine spoil restoration aimed at creating a multifunctional ecosystem in mine waste dumps. As we describe in the following sections, our strategy combines the simultaneous revival of traditional water harvesting systems, ground water recharge, enhanced biomass production and an adaptation to random recurrence of droughts in Rajasthan.

To our knowledge, the holistic strategy we suggest here has neither been proposed nor been implemented in its present from. However, our proposal incorporates lessons from different strategies tried in large number of places within Rajasthan and elsewhere. We suggest that the restoration strategy may be implemented as an experiment in adaptive restoration. In the following sub-sections we provide the details of the strategy proposed.

SITE LEVELING

To the extent possible, mine spoils need to be leveled or terraced in order to provide suitable substratum. Leveling will vary according to the type of mine, methods of mining and the way in which a particular area has been worked.

For instance, in case of surface and opencast mines the procedures will be leveling and fencing of the area. In case of shaft and underground mines although overburden can be treated in similar fashion but mined out areas and abandoned mines will have to have different strategy depending upon the context.

Mined out areas in hillside slopes may require contour dikes. Leveling will provide a base of coarse material over which to spread sediment. Some of the large mining pits that have developed into reservoirs can be developed as water-bodies aesthetically appealing for ecotourism and simultaneous fish culture to support local livelihoods.

Sediment Removal from Traditional Water Harvesting Systems

The main physical problems with mine spoils are shallow substrate of soil (or often lack of it), large cavities in the very coarse-grained substrate, very high stone content, extremely coarse texture, compaction, and the limited availability of moisture. The main chemical problems are the lack of nitrogen and phosphorus due to the lack of organic-matter content, low cation exchange capacity, and base saturation.

To overcome these challenges, addition of organic wastes is useful that ultimately increase N fertility and stimulates microbial action. Excavated sediment of ponds and tanks is an effective indigenous soil amendment practice in India. Pond silt is not only productive but also a seed bank for a variety of grasses, herbs, shrubs, and trees.

This silt, rich in organic material, can be used for preparation of topsoil layer of about 30-50 cm over the mine waste and leveled pits. The silt layer increases the productivity of the land and also helps in ground water recharge.

Transporting the silt away from ponds and using it for organic enrichment of mine spoil serves other purposes as well including the safe disposal of excavated sediment and solid waste, ecological restoration of mine-waste, and increased rainwater storage capacity for people and nature in village tanks. In addition to the above activity, *in situ* moisture conservation to encourage growth of vegetation over mine spoils could be useful.

For example, rehabilitation success to revegetate mine spoils in arid regions in India was achieved using a combination of *in situ* rainwater harvesting, soil amendments, and establishment of trees, shrubs and grasses. In many cases toxicity of mine spoils due to presence of metals affects the restoration plan.

In such cases, one of the methods that also can be applied along with sediment use is microbial enrichment of the ore where ever feasible. Some arbuscular mycorrhizal fungi (AM fungi) native to limestone mine spoils may play a critical role in rehabilitation of mine spoils in arid Rajasthan. AM-fungi have positive role in improving the water and nutrient uptake and enabling the plants to withstand high temperatures.

It should be noted here that the best practice should have been the safe storage of topsoil for reuse in restoration before mining commenced. This is a mandatory requirement. However, as that has not happened the strategy we propose is a practical one. How feasible is the proposal to remove sediment from water harvesting systems? Only a small number of the 48,000 ponds and tanks in Rajasthan have been periodically de-silted by villagers in the past. This neglect has resulted in large quantities of silt accumulation which can be removed from the water impounding area. For instance, de-silting of lake Man Sagar in Jaipur yielded approximately 0. 5 million cubic meters of silt which has been used for strengthening of embankment and construction of islands for migratory birds in their wintering grounds. Preliminary results of direct seeding in the islands have been very encouraging.

DIRECT SEEDING

Mine spoil treated with sediment can subsequently be fenced as protection againt grazing by free-ranging livestock and directly seeded with suitable species. Direct seeding of native species has been found to be is a useful and cost-effective restoration method globally. Direct sowing provides a large base for choice of species. Seed mixture for direct seeding must be carefully selected based on physical and chemical properties of mine spoil as well as ecological, economic and social criteria.

An useful approach is to include a set of selected indigenous herbs, grasses, shrubs and trees known as framework species because they help re-establish a basic forest structure that catalyses the recovery of biodiversity. Identification of a candidate set of framework tree species for direct sowing is guided by six main criteria:

- Multiple use local species that yield forest products to support livelihoods of the communities;
- Reasonably fast growth with dense spreading canopies which rapidly shade out weeds;
- Ease of collection and storage of seeds;
- A reasonable success in direct seeding;
- Attraction to wildlife (such as fruit and nectar or perching and nesting sites), because birds attracted by such species disperse seeds of other species into the restoration sites;
- Leguminous species, because they enhance soil microbial biomass and N mineralization and promote growth of other saplings growing in their vicinity.

As several keystone species are also socio-culturally valued, their inclusion in ecological restoration programme is helpful. Examples of such species are *Ficus religiosa* and *Ficus bengalensis, Bombax malabaricum, Prosopis cineraria,* and *Acacia* species. Fruit-eating animals and birds prefer to eat figs even when other food is abundant, as high calcium levels contribute to the desirability of figs as food for many animal species. Direct sowing such ecological and cultural keystone species may therefore enhance natural seed dispersal and accelerate succession in mine spoils. Some of the species may grow better in mine spoils.

A study by Rao and Tarafdar found that *Prosopis juliflora, Salvadora oleoides, Cenchrus ciliaris* can grow well in mine spoils. *Salvadora oleoides, Colophospermum mopane* and *Pithecellobium* dulce have been noted to be calcium-loving plants. These species can be useful for rehabilitation of gypsum mine spoil in Thar desert. Production of seedlings in the forest nursery requires large inputs in terms of time and money. The expenditure can be minimized, by opting for the direct sowing. Direct sowing is also comparatively easier in term of maintaining the proper proportion of species. It can be combined with planting and natural regeneration. Direct sowing helps in enhancing biodiversity per unit area, perhaps because it accelerates natural plant succession processes, as the ground cover created by newly germinated seeds acts as a nurse crop and can trap air-borne seeds from the vicinity.

Direct sowing requires simple technique for *in situ* collection of rainwater with the help of suitable soil work such as trenches and saucers. Thus, multi-tier vegetation (i. e. vegetation assemblage layers of herbs, shrubs and trees with differential height profiles) can be effectively developed.

Restoration and Livelihoods Improvement

The restoration strategy we suggest can work only in a conducive environment by providing "spaces" to local people to participate and take decision on different aspects of restoration.

This is the most crucial aspect to involve local communities and help them craft regimes for livelihoods improvement. Although there are administrative departments within the government addressing water, forests, mining and drought policies but there remains a lack of coherence in working of these. But, we can find "spaces" in existing legal and policy frameworks such as joint forest management.

We suggest that village forest management and protection committees created for joint forest management (JFM) can be employed for restoration of mine spoils and revival of rainwater harvesting systems provided that the government departments facilitate them and provide an opportunity to take their context-specific decisions.

A large network of forest protection committees is already functioning in India. An estimated 84,632 JFM groups involving 8. 38 million families are managing 17. 33 million hectares (mha) of forests (or 22. 2% of the 76. 5 mha of total recorded forest land) in 27 States of India. Total number of population covered under the community forestry programme is 62. 39 million (Bahuguna et al. 2004).

Likewise, there are 3667 village forest protection and management committees managing 3,76,766 ha plantations and forests in Rajasthan. These village institutions should be viewed as starting points for implementation.

We recognize, however, that all may not be well with JFM institutions and we need identification of challenges to craft workable local regimes that address restoration and livelihoods simultaneously. The real challenge lies in providing flexibility within the JFM policy to adapt and innovate for achievement of mine spoil restoration and revival of water harvesting systems.

FINANCIAL RESOURCES FOR RESTORATION

We must note here that dredging and transport of sediment and subsequent use for restoration shall be a costly affair. But society has limited options: either keeping mine lands derelict and ponds ruined, or enhancing productivity through restoration and revival.

The model proposed here may need resources ranging between US$5000 and US$10000 per ha depending upon the degree of mining and dereliction of lands. These are tentative estimates and a more robust idea of costs would become known after implementation of our strategy. In the year 1986 an economic analysis showed that costs for reclaiming mining overburden were INR 66000/ha.

The high costs, however, should not be surprising as a recent global analysis notes that annual cost in some cases can be as high as US$ 1 million per km^2 in programmes that require restoration to recover conservation value.

The resources can be mobilized by three stakeholder-departments: Mines, Water/Irrigation and Forest. Finances for silt removal can be mobilized from the ongoing efforts of the government for promotion of rainwater harvesting. Transport and spreading of sediment and protective fencing of the restoration areas can be financed by Department of Mines in collaboration with mine owners.

Forest Department may provide technical guidance and genetic resource (seeds, vegetative cuttings, plants). In addition, it would be useful to consider mobilizing financial resources for restoration through a clear legal framework that forces mine owners to finance restoration of mine spoils.

Along with command and control measures, it is also important to make use of financial tools like performance bonds. Government policies need to craft mechanism for mine owners to ensure that they are compensating for the environmental damage. Although design and enforcement of mine spoil restoration policy must accept the role of incentives to retrieve and store topsoil before starting the mining operations, but we note that such operation are possible before the start of mining only. It would be useful to clarify to mine owners that they need to finance the full cost related to:

- Pre-mining vegetation and soil removal,
- Protection to adjacent vegetation and trees as seed source,
- Post-mining restoration of overburden,
- Treatment to adjacent farmlands and streams affected by mining operations
- Cost of the preventive measures to mine-induced groundwater pollution.

These are essential measures that will provide robust incentives to mine owners for minimizing damage to the environment.

PILOT IMPLEMENTATION AT SARISKA TIGER RESERVE

The strategy we propose is conceptually robust as well as coherent with natural resource policies and can be implemented as such. Nonetheless, we realise the limitation of resource availability for large scale implementation. Thus, strategy can be implemented in a phased manner as a learning experiment initially in 215 closed mines around Sariska Tiger Reserve.

There have been mining problems primarily for dolomite in the reserve. A survey in 1989 found that over 200 mines fell in the protected forest area and over 40 in the partly protected area. Our suggestions for implementing the restoration around Sariska Tiger Reserve are guided by three core considerations:

- The reserve provides home to the western most population of *Panthera tigris tigris* globally. Even as the park is so critical for survival of tiger population there are severe conflicts over the use of natural resources. In early 1990s the Supreme Court of India ordered closure of 215 mines in and around Sariska Tiger Reserve. Restoration of these mines is expected to help in eco-development and provide learning and adaptation for large-scale implementation elsewhere. Adjoining district Jaipur has another potential wildlife habitat in Jamwa Ramgarh wildlife sanctuary. This protected area is a critical watershed forest for local communities. It may also be a spillover area for young tigers of the Sariska reserve. Anecdotal evidence suggests that mining for talc and marble may have degraded about 32 square kilometers area in the sanctuary.
- There are numerous traditional water harvesting systems surrounding Sariska Tiger Reserve. Although many of these have been revived, and new ones are being constructed, there is a critical need to address the availability of water for wildlife and people. Sediment removal from the traditional *johads* (ponds) shall be useful as an adaptation to random recurrence of drought in the area.
- In spite of conflicts over the use of natural resources and damage to crops and livestock by wild animals, local people still have a positive attitude towards the Sariska Tiger Reserve, because of tangible benefits derived from the Reserve in terms of fodder and fuelwood, wildlife tourism, and cultural and religious attitudes towards wild animals. Implementation of the strategy shall provide insights and help learn the operational aspects of integrated natural resource management, which is recognised in principle but seldom implemented.

Interdisciplinary approaches that accord respect to alternative knowledge systems are needed to address the effects as well as responses of human activities on tropical ecosystems. Restoration should thus focus on scientific understanding as well as strategic inclusion of local people, and their values and knowledge into the restoration process. In consonance, we have argued a strong case, and designed a commensurate strategy, for simultaneous revival of traditional water harvesting systems and restoration of mine spoils in Rajasthan.

In order to convert mine wastes, and ruined traditional ponds and tanks from environmental hazards to a learning opportunity, we need to implement a strategy for restoration, adaptation and productivity enhancement in such a manner that it also promises better prospects for sustainable livelihoods improvement to people dependent on natural resources.

Ancillary benefits of the proposed strategy include revival of traditional rainwater harvesting, enhancement of groundwater replenishment, safe disposal and use of solid waste, enhancement of the rainwater storage capacity in traditional village ponds and lakes, and drought-proofing in Rajasthan.

Restoration also promises to enhance the aesthetic appeal of blighted hills in a state noted for tourism. Although enough is known to initiate implementation, we recognize two issues for further research. First, collection of data regarding the precise area under various mining leases, active mining, waste dumps and abandoned mines is urgently required.

Second, as the implementation progresses, research and learning about the impacts of mine spoil restoration and concurrent revival of traditional water harvesting systems on livelihoods improvement shall be useful. Finally, we note that although our analysis has focused locally, lessons learnt through implementation may hold promise for other regions and countries facing similar challenges.

Chapter 9

Ground Water Contamination

Visibility in the national parks primarily involves a preservation issue instead of one of amelioration. Historically, Americans have placed a high value on good visibility, that is, the ability to see distant objects clearly. This appreciation of atmospheric visual clarity is evidenced in the country's early literature and art, including the journals of Lewis and Clark as well as the masterpieces of the great American landscape artists of the nineteenth century.

Today that yearning is demonstrated not only by the millions who flock each year to our western parks, but also in the high prices brought by those artists works of a century ago and by the interest in Ansel Adams's simple, yet dramatically clear, black-and-white photographs of Yosemite and other wonders.

Over the past one hundred years, Congress has acted to preserve many of the nation's natural wonders. It did so by creating and by continually expanding the national parks, wilderness areas, monuments, recreation areas, and wild and scenic rivers. Since the 1950s, there has been an increasing concern that this beauty is threatened by industrial development and population growth.

Pollution from coal-fired power plants became a special concern in 1963 with the advent of the first unit of the Four Corners Power Plant near Farmington, New Mexico. It produced a plume that could be seen clearly for many miles, reducing the clarity of the visual experience in areas of northwestern New Mexico, southeastern Utah, southwestern Colorado, and northeastern Arizona.

By the later 1960s and the early 1970s, smog began to appear in Yosemite Valley, California, on warm summer days. Battles erupted over the visibility effects of proposed coal-fired power plants on the Kaiparowits Plateau and near Capitol Reef National Park, both in southern Utah.

The increased publicity and concern resulted in magazine and newspaper articles decrying the loss of visual clarity, particularly in the western United States, and precipitated political pressures in

Congress for legislative steps to protect visibility. Those pressures culminated in the August 1977 adoption by Congress of the nation's first specific visibilityprotection requirements for national parks and national wilderness areas. One of the large issues raised by these developments is whether the value of visibility protection outweighs the cost, including both air pollutioncontrol equipment and the regulatory system.

The study reported in this section was designed to improve our ability to measure the benefits of visibility and to provide some preliminary estimates of the value of that visibility in several major national parks and for the region in which they are located. Visibility is the ability to see both colour and detail over long distances. Human perception of visual air quality is associated with the apparent colour contrast of distant visual targets.

As contrast is reduced, a scene "washes out" both in terms of colour and in one's ability to see distant detail. What, then, is the nature of the preservation value of visibility? That value has at least two possible components. First, a scenic resource such as the Grand Canyon attracts large numbers of recreationists. The quality of the experience of these individuals depends in great part on air quality, in that scenic vistas are an integral part of the Grand Canyon "experience. "

Accordingly, the air quality at the Grand Canyon is valuable to recreationists. We might call this economic value, or willingness to pay by users for air quality at the Grand Canyon, user value. Thus, recreationists in the national parklands of the Southwest should be willing to pay some amount to preserve air quality for each day of their own use if their recreational experience is improved or maintained by good air quality.

The second component of preservation value we have termed existence value. Individuals and households who may never visit the Grand Canyon may still value visibility there simply because they wish to preserve a national treasure. Visitors also may wish to know that the Grand Canyon retains relatively pristine air quality even on days when they are not visiting the park. Concern over preserving the Grand Canyon may be just as intense in New York or Chicago as it is in nearby states and communities. Thus, preservation value has two additive components, user value and existence value.

During the summer of 1980, more than 600 people in Denver, Los Angeles, Albuquerque, and Chicago were shown five sets of photographs depicting both clear conditions and regional haze. Each set consisted of five photographs of a national park vista with different visual air quality of a general nature, that is, generally increased haziness. The vistas shown were from Grand Canyon, Mesa Verde, and Zion. Summer was chosen for the survey because it is the season of peak visitation.

These photographs were placed on display boards as full-frame, 8-×10-inch textured prints, arranged from left to right in ascending order of visual air quality, with each vista in a separate row. The participants were asked how much they would be willing to pay for visibility as shown in the five sets of photographs, from worst to best. They were also asked how much they would be willing to pay to prevent a plume from being seen in a pristine area.

Two photographs were used in this connection, one with and the other without a plume. The photographs were taken from Grand Canyon National Park at the Hopi fire tower observation point and toward Trumbull Mountain. They were both taken at 9:00 AM, so the lighting on the canyon wall and other features is the same. Both photographs have the same light, high cirrus cloud layer in the sky. The plume is a narrow gray band crossing the entire vista in the sky, except where it appears in front of the top of Trumbull Mountain.

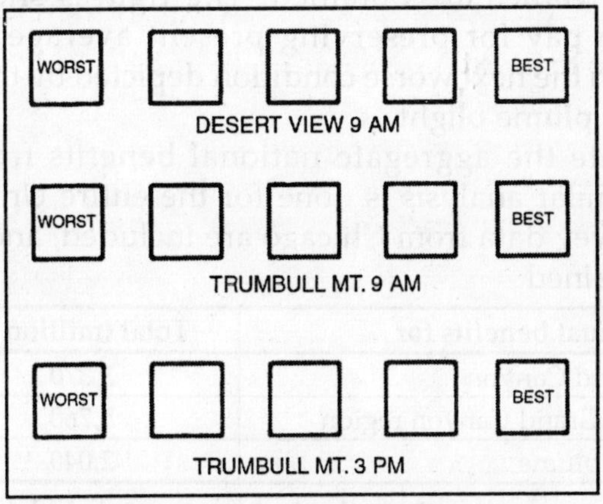

Fig. Grand Canyon Photograph Board.

The source was not industrial or municipal pollution, but a controlled burn in the area around the Grand Canyon. However, the effect was comparable to what a large industrial source might produce.

The bidding game based on these photographs reveals the household's willingness to pay for preserving or improving the degree of visibility in specific locations of the national park area described earlier. The bids offered by respondents in the aggregate preservation-value section of the survey encompass both pure existence value and users' valuations of preserving visibility. Since the results did not permit a completely clean distinction between the two types of bids, further discussion will concentrate on the preservation-value section of the survey.

The benefits derived from the interview results can be extrapolated to populations larger than that in the sample (the sample was chosen in as random a manner as practical) by applying statistical techniques to the results of the survey. The amount of bids offered by interviewees to preserve or improve visibility is related to such factors as income, education, and other personal characteristics.

These relationships can be quantified using regression analysis. After this is done, it is possible to estimate the value of benefits to residents of the whole Southwest region as well as the entire nation. This is done by substituting the average values for these characteristics for each state into the relationship established by the regression analysis and calculating what the average value of the bid of a person in that state would be. This value can then be multiplied by the population of the state as a whole to get a total bid.

When the analysis is performed for the southwestern United States, the following values are obtained. The figures show the annual willingness to pay for preserving present average conditions, as contrasted with the next worse condition depicted by the pictures, and for preventing plume blight.

To estimate the aggregate national benefits from preserving visibility, a similar analysis is done for the entire United States, but additional survey data from Chicago are included, and the following values are obtained.

Annual benefits for	**Total (million $)**
Grand Canyon	3,370
The Grand Canyon region	5,760
The plume	2,040

These figures, even though their accuracy is highly uncertain, imply that very large existence values characterize the areas in question. However, some recent and highly preliminary experiments with surveys imply that these figures may be much too high.

Several other observations on the outcomes of the analysis of the actual interview results are worth mentioning here. First, in the conventional view of the demand for environmental quality, there is a smooth tradeoff between higher successive levels of environmental quality and economic benefits, with successive units commanding less incremental willingness to pay. The survey respondents, however, placed a much higher value on a small initial diminution in visual clarity than on comparable subsequent decreases.

Second, and again somewhat contrary to expectations, neither past nor prospective visits to the Grand Canyon region were shown to be important determinants of preservation value. On the average, those who had never seen the Grand Canyon valued it as highly as those who had. Third, and again unexpectedly, a household's distance from the region had no significant relationship to the size of its bids.

When corrected for income and other differences, people in Chicago bid fully as high as those closer by. However, preliminary further investigation suggests that this result may not be very robust, being sensitive, for example, to the sequence in which people are asked about their valuation of various public goods. Because the Grand Canyon is the dominant feature in a region with many visitor attractions, one must be especially cautious in extending these preliminary findings to other recreational attractions.

It seems likely that there are only a very few natural phenomena in the United States about which Americans have such strong feelings. Obvious candidates for this short list would be Old Faithful (in Yellowstone National Park), Niagara Falls, and perhaps a few others.

The main conclusion of this visibility study is that the magnitude of the annual benefits, when aggregated across households, is impressive. While these are necessarily rather crude extrapolations, the survey results suggest that Americans place great value on the preservation of air quality in the Grand Canyon region, and that this valuation is not necessarily localized in the Southwest. Further, the survey results suggest that pure existence value may overwhelm a substantial user value for the national parks in the region.

ACID RAIN

Acid rain and other mechanisms for the dispersion and deposition of acid formed from sulfur and nitrogen emitted from various sources are complex and ill-understood phenomena. In addition, methods for estimating the economic losses resulting from damages or economic benefits of prevention of acid rain are not well developed, nor was it possible within the scope of the project described here to make much progress in developing them.

Consequently, since the estimates of benefits made for controlling acid rain are very crude and of no particular interest in terms of methods development, the discussion here will be very brief. It is included primarily because of current intense interest in the phenomenon, and because the analysis that was done provides some guidance concerning directions for future research.

Acid deposition, among all the areas covered in this volume, is perhaps the one most crying out for additional methods development and improved estimates. Let us turn first to its possible effects on agriculture and forestry.

Increases in soil acidity can negatively affect the yields of certain field crops, but this can be offset by modestly increasing liming operations which already occur for acid-sensitive crops. Therefore, the benefits of controlling acid rain for this purpose would be small. It is known that direct damage to the plant can occur from acid deposits on leaves, flowers, and fruits, but there is virtually no basis for estimating the amount of such an effect.

Forest growth can be affected in a similar fashion in that there can be both indirect effects, through the soil, as well as direct effects. As in the case of certain field crops, there may even be short-run favorable effects, for example, when the acid dissolves plant nutrients and makes them more available to the trees. Nonetheless, in the longer term this could result in reduced soil fertility and slower tree growth. If some strong assumptions are made, an estimate can be made of damages resulting from retarded growth.

If one assumes—and there is some evidence pointing in this direction from Swedish studies—that acid rain would reduce timber growth in Minnesota and east of the Mississippi (the area of the country thought to be most affected by acid) by 5 per cent annually, the reductions in yield would decrease the worth of timber production about $600 million per year.

Assuming that other services of forests, such as watershed protection, fishing, and hunting, also were to be reduced by 5 per cent, and based on crude estimates by others of the possible overall value of these services, the total damage including timber and other services might come to about $1. 75 billion. This is a substantial sum, but not very large relative to the costs of controlling acid deposition. There might also be effects on human health, say, by the acid dissolving and mobilizing heavy metals so that larger concentrations would get into drinking water or the human food chain.

The present state of knowledge does not permit even very crude estimates to be made of this possibility. Higher acidity in municipal and industrial water systems might also result in increased corrosion in piping, appliances, and cooling systems. But the diminishing acidity in such systems, by the use of lime, is a routine operation and can be accomplished at small cost.

The big danger in watercourses appears to be those features of the aquatic ecosystem itself which mankind values. Acid conditions in a watercourse tend to destroy the small plants and animals (plankton), that are the initial links in the fish food chain, and this has a negative effect on fish population. But the primary way in which fish populations are destroyed is different. As noted earlier, acid in water bodies tends to mobilize toxic metals and increase their concentration in the water.

The reproductive capacity of many species of animals, including fish, is adversely affected by the presence of excessive amounts of toxic metals. Thus, for a time, as fish numbers decline, the ones that remain increase in size as competition for food declines, but then rather abruptly there are none left. This, of course, destroys commercial and recreational fisheries. The value of fish taken by commercial freshwater fishing in the United States is not very large, so the loss there, at least as measured by present market prices, would not be very great. On the other hand, the value of freshwater recreational fisheries is relatively enormous. Let us make the extreme assumption that all such fisheries in Minnesota and other areas east of the Mississippi would totally disappear. If we then take estimates of willingness to pay for fishing from other studies, it appears that the loss could, at an outside limit, be on the order of $10 billion per year in 1979 prices. Additional losses would be caused by the decline of terrestrial and aquatic animals (other than fish) who are partly or wholly dependent on the aquatic food chain-certain species of water fowl, for example.

The other area where our study suggests really major damages might occur is deleterious effects on materials. As noted previously, acid corrodes metals, eats away limestone, is harmful to paints and other coatings and finishes, and damages cloth. Given the huge number of such items which exist and are exposed to the atmosphere, it is not very surprising that benefits from protecting them might be large.

Again, in Sweden, where the problems of acid rain first received widespread attention (because of prevailing winds, Sweden gets inputs of sulfur and nitrogen compounds from the Ruhr, the Rotterdam petrochemical complex, and Great Britain), a study has been made of per capita damages of corrosion and soiling.

If one makes the assumption, once again very gross, that this same estimate can be applied to all persons dwelling in Minnesota and east of the Mississippi, one gets an annual benefit from avoiding acid rain of about $14 billion. Putting together the various dollar estimates, the benefits of preventing acid rain in 1978 dollars amounts to nearly $26 billion—a hefty amount indeed.

But as stated, many extreme assumptions were made in generating these numbers, and they are no doubt too high by far. An educated guess by the research team was that the actual figure is probably not more than $5 billion for a condition that is characterized by severe effects on the entire eastern United States.

Of course this number cannot be taken very seriously, because even if it were correct, in all other respects it neglects the large adjustments in demand and supply which would accompany the types of changes contemplated. An approach much more like that described for the agriculture study in the South Coast Basin would be appropriate in a more sophisticated study. Where are the greatest potential benefits of protection from acid rain likely to lie? These are, in rough order, materials damage, aquatic ecosystems effects, and effects on agriculture and forestry. These categories of damages certainly merit further study.

But our progress in economic research on these questions is highly dependent on improved dose-response relationships. The relationship between the intrusion of acid into a watercourse and its ultimate effect appears to be extraordinarily complex.

For example, what sudden changes could occur after a previously uneventful period, and what difficulties would be encountered in reversing them after they have occurred? Indeed, could they be reversed completely? Ecologists place great value on diversity of species as an indicator of a healthy, stable ecological system. Acidification of streams is known to reduce diversity. But it is not well understood how this ultimately affects characteristics of the stream that we value. This problem seems ripe for joint work between economists and ecologists.

GROUNDWATER CONTAMINATION

While the extent of groundwater contamination is not accurately known, it is thought to be widespread and is the focus of much public apprehension. Contaminants in groundwater range across an enormous list of chemical substances, and usually no thorough checks for contamination are made until there is reason to suspect a problem. Even at extremely low concentrations, many toxic chemicals pose serious, irreversible, health risks.

In many of the cases investigated, well water has been found to contain concentrations above, and often several orders of magnitude higher than, those commonly encountered in raw or treated drinking water drawn from contaminated surface sources. Thus, while water from most wells no doubt is safe, the widespread nature of the contamination and its potential seriousness merit the public attention the problem is getting. The intent of the case study in this chapter is to develop methods for estimating benefits from preventing contamination of groundwater-based drinking-water supplies.

This, so far as I know, is the first study to attempt to quantify such benefits. As in the studies discussed in other chapters, the quantitative results reported here must be regarded as largely experimental, but the numbers turn out to be impressively large. For any chemical source, the extent of groundwater contamination is determined by the characteristics of the underground storage medium-called an aquifer. Groundwater in shallow, alluvial aquifers, typically moves less than a foot per day. That flow is governed by recharge and discharge rates from the aquifer, and by the aquifer's permeability.

Contaminants are transported by diffusion together with the slow underground flow of groundwater. In that oxygen-poor environment, chemical or physical processes of contaminant degradation proceed very slowly. Thus the contaminant plume may move great distances, with hardly a change in toxicity levels, and may therefore reach drinkingwater wells. Among the principal sources of groundwater contamination are wastedisposal landfills and impoundments, accidental pills of chemical substances, and abandoned oil and gas wells.

Most groundwater contamination can be traced to chemicals leaching into the aquifer from poorly constructed and managed industrial or municipal landfills, surface impoundments, or outright illegal dumps. Contamination from such sources has often been in process for years, and sometimes for decades. To date, most groundwater contamination incidents have been discovered only after a drinking-water source has been affected.

By the time suspected aquifer contamination is verified in samples drawn from drinking-water wells, the problem may be irreversible. Stricter regulation of the disposal of potential contaminants in other environmental media, particularly air and surface waters, and the consequent rising cost of such disposal, is likely to increase the flow of wastes to land disposal and aggravate the threat to groundwater.

Benefit analysis of controlling groundwater contamination requires, as usual, quantification of several link between sources and receptors. One must know the location and strength of actual or potential sources of contamination, and must be able to model the spread of the contaminant plume in the aquifer. One must know the numbers of persons exposed to contaminated groundwater and the extent and timing of their exposures. One must know the "dose-response relationship," the nature and extent of health effects on the population at risk.

And finally, one needs a way of converting health effects into monetary, or dollar, values. We are far from being able to quantify these link between sources and receptors with precision. In each case, there is a need for substantially improved methods and data. Actually, while it is referred to as a landfill, this is a rather euphemistic term—dump would be a better word—but I shall stick with the conventional usage.

Price applied to the New Jersey Department of Environmental Protection for a license to conduct a sanitary landfill operation. The application listed the materials that Price intended to accept at the landfill, and specifically excluded "Chemicals." He was issued a certificate authorizing operation of a solid waste disposal facility. Authorities inspected the landfill, citing Price for accepting chemical wastes and formally advising him of the violation. Nonetheless, Price continued accepting significant quantities of chemical wastes.

After that date, no chemical wastes were disposed of at the landfill, although it continued in operation. Price terminated the landfill operation and covered the site with fill material. The site has not been used since then. Price accepted approximately 9 million gallons of toxic and flammable chemical and liquid wastes, either in drums or directly into the ground. Among others, these included glycolic, nitric, and sulfuric acid, caustics and spent caustic wastes, cesspool waste, chemical resins and other waste chemicals, chloroform, and cleaning solvents.

Price's Landfill is situated over the Cohansey aquifer, the principal source of Atlantic City's water supply, and the separation between landfill and aquifer is a relatively permeable layer. Waste from the landfill is free to leach into the aquifer; the direction of flow in the aquifer is eastward, toward Atlantic City's wells.

Chemicals in the leachate therefore can be carried into the private and public water-supply wells, and people can be exposed to these chemicals in drinking water. Test wells drilled near the landfill by the EPA show that groundwater in the aquifer is contaminated and that the plume of contamination indeed is moving toward Atlantic City's wells.

But estimation of actual or potential human exposures requires either considerable information on, or heroic assumptions about, the mechanism by which toxics are transported from the source of contamination to the water-supply wells. This is the second linkage mentioned earlier. Shortly it will be clear why discussion of this linkage logically precedes the first quantification of the source of the contamination.

Efforts to understand and model the source to receptor

linkages, called groundwater solute transport, are relatively recent. While there has been considerable earlier work on salinity transport, study of the more difficult cases of chemically reactive toxic groundwater contaminants is less advanced. Improvements in our ability to model these phenomena must be a prime objective for future research.

For purposes of analyzing the Price's landfill situation, the researchers chose and estimated numerically a technique called the Wilson-Miller solute-transport model. This relatively simple model was chosen because of time and funding limitations for the research. The model chosen does appear to fit the Price's landfill situation relatively well and has been judged adequate for conducting this experiment. Future research should determine whether more complex models yield substantially different results.

But to apply any solute transport model, it is first necessary to have so-called source-term information, that is, the amounts of material entering groundwater and their distribution over time. Much of the activity at Price's landfill was illegal. It therefore seems unlikely, to say the least, that careful records of what went into the pit were kept. Indeed, for a large number of chemical substances there is no information at all about the amounts that have been dumped there.

Where such records exist, or if leaching rates are known or can be calculated, deliveries of pollutants to the aquifer can be estimated directly. In the situation exemplified by Price's landfill, which is typical of much groundwater contamination, there is only one way to estimate the quantity. Since we have information on what is already present in test wells drilled by the EPA, the solute-transport model can be run "backwards," so to speak, and used to infer what amount there had to be to produce the existing groundwater concentrations. This is why, logically, the discussion of the transport model precedes discussion of the source term.

The reader should be cautioned that this estimate, while necessary, is based on many assumptions and involves great uncertainty. Just to give one example, the procedure assumes that releases occur at a constant rate. This may not be true for some pollutants, and "slugs" may be released which cause transients of pollution in much higher concentrations than would be predicted by the model. But given the computed source term, the model can be run "forward" to compute concentrations at any well drawing on the aquifer—the production wells of Atlantic City, for example—and for any time after some contaminant enters the aquifer.

Those concentrations, and the times at which they are projected to occur, were computed for the wells from which Atlantic City's Municipal Water Authority pumps its water. Assuming that no mitigating action is taken, this provides the link that specifies the exposure of the population to contamination from Price's landfill.

To take the next step, one must have dose-response information—that is, the actual health risk stemming from the contamination. To make this link, information published by the EPA was used. There is a section of the Clean Water Act that requires the EPA to estimate excess cancer risks for 129 chemicals called *priority pollutants.*

Many of these priority pollutants are leaching from Price's landfill into the Cohansey aquifer. Using this information, the probability of excess mortality from cancer was estimated for the population of Atlantic City. While this procedure is the best available based on existing information, the reader should be aware that, for this purpose, the risk factors provided by the EPA are both incomplete and very uncertain.

For example, there are many pollutants that have been identified in groundwater that are not on the EPA list, and extrapolations from animal toxicity tests to human risks are quite uncertain. Additionally, it is assumed that each chemical risk is independent of each other chemical risk so that risks can simply be added up across chemical categories.

It is well known that a synergistic effect can occur, making the combined toxicity of two chemicals greater than the sum of the effects of each one taken independently.

Again, with all these cautions in mind, I turn to the next, and final, step, the monetary evaluation of damages. The values ranged from $ 100,000 to $1 million per death. These were then multiplied by the mortality numbers calculated in the risk analysis to obtain the total benefit from averting the damage which would otherwise emanate from Price's landfill. The total benefit ranges from $180 million to $1. 8 billion.

Those are large amounts, and one must be clear about what they mean. Say that, at a site similar to Price's landfill, there is a comparable release of contaminants into a similar aquifer, and that the release goes unnoticed for two decades.

Chapter 10

Effects of Ozone Depletion

Ozone is both beneficial and harmful to us. Near the ground, ozone forming as a result of chemical reactions involving traffic pollution and sunlight may cause a number of respiratory problems, particularly for young children. However, high up in the atmosphere in a region known as the stratosphere, ozone filters out incoming radiation from the Sun in the cell-damaging ultraviolet (UV) part of the spectrum. Without this ozone layer, life on earth would not have evolved in the way it has.

Concentrations of ozone in the stratosphere fluctuate naturally in response to variations in weather conditions and amounts of energy being released from the Sun, and to major volcanic eruptions. Nevertheless, during the 1970s it was realised that man-made emissions of CFCs and other chemicals used in refrigeration, aerosols and cleansing agents may cause a significant destruction of ozone in the stratosphere, thereby letting through more of the harmful ultraviolet radiation.

Then in 1985 evidence of a large "ozone hole" was discovered above the continent of Antarctica during the springtime. This has reappeared annually, generally growing larger and deeper each year. More recently, fears have emerged about significant ozone depletion over the Arctic, closer to the more populous regions of the Northern Hemisphere.

In response to this and additional fears about more widespread global ozone depletion, the Montreal Protocol on Substances that Deplete the Ozone Layer was implemented in 1987. This legally binding international treaty called for participating developed nations to reduce the use of CFCs and other ozone depleting substances. In 1990 and again in 1992, subsequent Amendments to the Protocol brought forward the phase out date for CFCs for developed countries to 1995.

Protecting the ozone layer is essential. Ultraviolet radiation from the Sun can cause a variety of health problems in humans, including skin cancers, eye cataracts and a reduction in the body's immunity to disease. Furthermore, ultraviolet radiation can be damaging to microscopic life in the surface oceans which forms the basis of the world's marine food chain, certain varieties of crops including rice and soya, and polymers used in paints and clothing. A loss of ozone in the stratosphere may even affect the global climate.

International agreements and other legislation have gone a long way to safeguarding this life-supporting shield. Nevertheless, for there to be real and long-lasting success, everyone must become part of the solution. Individual efforts taken together can be powerful forces for environmental change. There are a number of things that we, as individuals, can do to both protect the ozone layer.

These include proper disposal of old refrigerators, the use of halon-free fire extinguishers and the recycling of foam and other non-disposable packaging. Finally, we should all be aware that whilst emissions of ozone depleters are now being controlled, the ozone layer is not likely to fully repair itself for several decades. Consequently, we should take precautions when exposing ourselves to the Sun.

ANTARCTICA

Why is the ozone hole over Antarctica? That is one of the first questions that comes to mind when people think about the ozone hole. Every winter and spring since the late 1970s, an ozone hole has formed in the stratosphere above the Antarctic continent. In recent years this hole has become both larger and deeper, in the sense that more and more ozone is being destroyed. As summer approaches, the hole repairs itself, only to reform during the following spring.

Man-made emissions of ozone-depleting chemicals (ODCs) occur mainly in the Northern Hemisphere, with about 90% released in Europe, Russia, Japan, and North America. The global system of winds helps to distribute these ODCs, moving them towards the North and South Poles. Normally, the chlorine and bromine which destroys the ozone when free, is locked up within the fairly unreactive and stable ODCs.

However, during the Antarctic winter months (June to August) when the region receives no sunlight, the stratosphere becomes cold enough (–80°C) for high level [ice] clouds to form, called Polar Stratospheric Clouds (PSCs). These PSCs provide an ideal catalytic surface on which chlorine and bromine can react with the ozone, thus destroying the ozone layer.

This reaction requires sunlight, and therefore only begins when the Sun returns to Antarctica in spring (September to October), before the PSCs have had a chance to melt. The ozone hole disappears again when the Antarctic air warms up enough during late spring and summer

During the Southern Hemisphere winter Antarctica is isolated from the rest of the world by a natural circulation of wind called the polar vortex. This prevents atmospheric mixing of stratospheric ozone, thus contributing to the depletion of ozone above Antarctica. Although some ozone depletion occurs over the Arctic, meteorological conditions there are very different to Antarctica and so far have prevented the formation of ozone holes as large as in the Southern Hemisphere.

ARCTIC

Even a small depletion of stratospheric ozone in the Arctic region would give cause for considerable concern due to the higher populations in the higher latitudes of the Northern Hemisphere.

Arctic ozone depletion has not been as marked as over the Antarctic for two reasons. Firstly, stratospheric temperatures seldom fall below -80°C due to frequent exchange of air with the mid latitudes. Without these low temperatures, the formation of polar stratospheric clouds is precluded, and ozone destruction rates are much lower.

Secondly, the polar vortex of wind in the Arctic region is much weaker than the corresponding one in the Southern Hemisphere, and usually dissipates in late winter before sunlight returns to initiate a rapid ozone destruction. The differences between the two polar regions result in part from the larger land mass in the Northern Hemisphere, which causes more irregular airflow in the atmosphere.

Nevertheless, analysis of satellite data reveals that the loss of ozone in the Northern Hemisphere is now proceeding faster than previously thought. In 1989, NASA's Airborne Arctic Stratospheric Expedition, the first comprehensive research expedition to explore the Arctic region, found that the Arctic stratosphere in winter has almost as much chlorine monoxide as is found in Antarctica, the same destructive chlorine that causes the Antarctic ozone hole.

While no Arctic ozone losses comparable with those in the Antarctic have occurred, localised Arctic ozone losses have been observed in winter and spring concurrent with observation of elevated levels of reactive chlorine, made available through man-made emissions of CFCs and other ozone depleting chemicals (ODCs). Ozone losses have increased greatly during the 1990s in the Arctic.

The springtime ozone hole in the Arctic is expected to grow larger during the coming decades as a result of global warming, before recovering after 2020. Though greenhouse gases cause atmospheric warming at the Earth's surface, they cool the stratosphere by trapping more heat below, in the troposphere. Since ozone chemistry is very sensitive to temperature, particularly at -80°C when polar stratospheric clouds can form, this stratospheric cooling may result in more ozone depletion in the Arctic. Furthermore, computer models predict that temperature and wind changes induced by greenhouse gas emissions may allow a stronger and longer-lasting atmospheric vortex to form above the Arctic, as in the Antarctic, enhancing ozone depletion.

CAUSES OF OZONE DEPLETION

Ozone depletion occurs when the natural balance between the production and destruction of stratospheric ozone is tipped in favour of destruction. Although natural phenomena can cause temporary ozone loss, chlorine and bromine released from man-made compounds such as CFCs are now accepted as the main cause of this depletion.

It was first suggested by Drs. M. Molina and S. Rowland in 1974 that a man-made group of compounds known as the chlorofluorocarbons (CFCs) were likely to be the main source of ozone depletion. However, this idea was not taken seriously until the discovery of the ozone hole over Antarctica in 1985 by the British Antarctic Survey. Chlorofluorocarbons are not "washed" back to Earth by rain or destroyed in reactions with other chemicals. They simply do not break down in the lower atmosphere and they can remain in the atmosphere from 20 to 120 years or more. As a consequence of their relative stability, CFCs are instead transported into the stratosphere where they are eventually broken down by ultraviolet (UV) rays from the Sun, releasing free chlorine.

The chlorine becomes actively involved in the process of destruction of ozone. The net result is that two molecules of ozone are replaced by three of molecular oxygen, leaving the chlorine free to repeat the process:

$$Cl + O_3 \rightarrow ClO + O_2$$
$$ClO + O \rightarrow Cl + O_2$$

Ozone is converted to oxygen, leaving the chlorine atom free to repeat the process up to 100,000 times, resulting in a reduced level of ozone. Bromine compounds, or halons, can also destroy stratospheric ozone. Compounds containing chlorine and bromine from man-made compounds are known as industrial halocarbons.

Emissions of CFCs have accounted for roughly 80% of total stratospheric ozone depletion. Thankfully, the developed world has phased out the use of CFCs in response to international agreements to protect the ozone layer. However, because CFCs remain in the atmosphere so long, the ozone layer will not fully repair itself until at least the middle of the 21st century. Naturally occurring chlorine has the same effect on the ozone layer, but has a shorter life span in the atmosphere.

CHLOROFLUOROCARBONS

Chlorofluorocarbons or CFCs (also known as Freon) are non-toxic, non-flammable and non-carcinogenic. They contain fluorine atoms, carbon atoms and chlorine atoms. The 5 main CFCs include CFC-11 (trichlorofluoromethane - $CFCl_3$), CFC-12 (dichloro-difluoromethane - CF_2Cl_2), CFC-113 (trichloro-trifluoroethane - $C_2F_3Cl_3$), CFC-114 (dichloro-tetrfluoroethane - $C_2F_4Cl_2$), and CFC-115 (chloropentafluoroethane - C_2F_5Cl).

CFCs are widely used as coolants in refrigeration and air conditioners, as solvents in cleaners, particularly for electronic circuit boards, as a blowing agents in the production of foam (for example fire extinguishers), and as propellants in aerosols. Indeed, much of the modern lifestyle of the second half of the 20th century had been made possible by the use of CFCs. Man-made CFCs however, are the main cause of stratospheric ozone depletion. CFCs have a lifetime in the atmosphere of about 20 to 100 years, and consequently one free chlorine atom from a CFC molecule can do a lot of damage, destroying ozone molecules for a long time. Although emissions of CFCs around the developed world have largely ceased due to international control agreements, the damage to the stratospheric ozone layer will continue well into the 21st century. Before the Montreal Protocol, CFCs were used in a variety of industrial and commercial appliances.

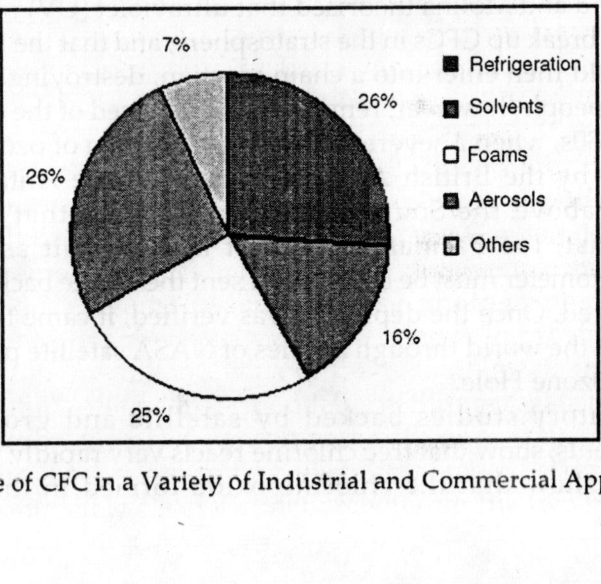

Fig. Use of CFC in a Variety of Industrial and Commercial Appliances

DOING OUR BIT FOR THE OZONE LAYER

There are a number of steps that we can all take both as individuals and as groups to protect the Earth's ozone layer. We have all been part of the ozone depletion problem, through the use of chemicals in everyday products. However, we can all be part of the solution.

Following the Montreal Protocol most ozone depleting chemicals (ODCs) have or are being phased out of use in most target applications such as aerosols, refrigeration and air conditioning. However, consumer products bought prior to international agreements may still be in use in our homes and offices and cannot easily be replaced.

Large appliances, such as refrigerators, have long lifetimes and early replacement would entail great cost. Proper care and maintenance of equipment to ensure that the CFCs they contain are never released to the atmosphere should be applied. Remember that a single CFC molecule can destroy 100,000 ozone molecules.

In addition, if purchasing fire extinguishers try to avoid any that contain halons, which have bromine in them. Purchase carbon dioxide, water, or dry chemical extinguishers instead. Finally, although foam packaging is CFC-free, some products contain HCFCs (hydrochlorofluorcarbons), which while far less damaging to the ozone layer, could contribute substantially to global warming. Avoid those that do. Use and re-use non-disposable packaging.

Evidence for Ozone Depletion

In 1974, after millions of tons of CFCs had been manufactured and sold, chemists F. Sherwood Rowland and Mario Molina of the University of California began to wonder where all these CFCs ended up. Rowland and Molina theorised that ultraviolet (UV) rays from the Sun would break up CFCs in the stratosphere, and that the free chlorine atoms would then enter into a chain reaction, destroying ozone.

Many people, however, remained unconvinced of the danger until the mid-1980s, when a severe springtime depletion of ozone was first monitored by the British Antarctic Survey above Antarctica. The depletion above the South Pole was so severe that the British geophysicist, Joe Farman, who first measured it assumed his spectrophotometer must be broken and sent the device back to England to be repaired. Once the depletion was verified, it came to be known throughout the world through a series of NASA satellite photos as the Antarctic Ozone Hole.

Laboratory studies backed by satellite and ground-based measurements, show that free chlorine reacts very rapidly with ozone. They also show that the chlorine oxide formed in that reaction undergoes further processes that regenerate the original chlorine, allowing the sequence to be repeated up to 100,000 times. This process is known as a "chain reaction". Similar reactions also take place between bromine and ozone. Observations of the Antarctic ozone hole have given a convincing and unmistakable demonstration of these processes.

Based on data collected since the 1950s, scientists have determined that ozone levels were relatively stable until the late 1970s. Severe depletion over the Antarctic has been occurring since 1979 and a general downturn in global ozone levels has been observed since the early 1980s.

Global ozone levels declined by about 3% between the late 1970s and early 1990s. This rate of decline is about three times faster than that recorded in the early 1970s. In addition to Antarctica, ozone depletion now affects almost all of North America, Europe, Russia, Australia, New Zealand, and a sizeable part of South America. Short-term losses of ozone can be much greater than the long-term average. In Canada, ozone depletion is usually greatest in the late winter and early spring, associated with the development of an Arctic ozone hole.

EYE DISORDERS

Strong UV radiation from the Sun can cause inflammation of the cornea, the outer coating of the eyeball, leading to photokeratosis or "snow blindness". Symptoms of this kind of infection include the reddening of the eyes, a sensitivity to light, enhanced excretion of tears, the feeling of having some dirt in one's eye, and pain. The trauma appears 3-12 hours after exposure. Thanks to the quick regeneration of the eye cells, symptoms will normally disappear within a few days. A long-term exposure to UV radiation however, may cause permanent damage to the cornea.

UV radiation also enhances the dimming of the eye's lens, which means that potential cataracts begin to evolve at earlier ages. A cataract is a partial or complete opacity of the lens of the eye and the largest cause of blindness in the world.

Part of the UV radiation reaches the back of the eye, causing cells in the retina to slowly begin to deteriorate. Damage will in time particularly occur to near vision. If not operated upon blindness can occur. Radiation is partly absorbed in the lens of an adult eye, but will go right through the lens of a child, reaching the back of the eye. For this reason, children's eyes in particular should be protected against strong sunlight.

Other common eye diseases associated with increased UV

radiation are eye cancer, conjunctivitis and pterygium. Conjunctivitis is an inflammation of the membrane covering the anterior portion of the eyeball. Pterygium is a thickening of the membrane that covers the eyeball.

A loss of ozone in the stratosphere will increase the amount of UV radiation reaching the Earth's surface. If stratospheric ozone decreases by 10% during the spring and summer, the UV radiation dose increases by about 12%. Unlike the skin, which can adapt to UV radiation by becoming browner and thicker, the eye does not have any such defence mechanisms.

On the contrary, research shows that eyes become more sensitive with increased exposure to UV radiation. Increased exposure to UV radiation from ozone depletion is expected to increase the number of people experiencing cataracts and other eye disorders. A 1% decrease in stratospheric ozone may result in 100,000 to 150,000 additional cases of blindness due to eye cataracts worldwide.

GLOBAL CLIMATE

Increases in solar UV radiation as a result of ozone depletion could affect terrestrial and aquatic biogeochemical cycles, thereby altering both sources and removal sinks of greenhouse gases, for example carbon dioxide, and possibly other trace gases including man-made pollutants. Likely effects of this may include an increase in air pollution in urban centres, and acid rain in rural areas.

Whilst increases of UV radiation may affect the production and removal of carbon dioxide, the main greenhouse gas, ozone depletion itself can influence the global climate. Ozone is also a greenhouse gas, and as well as filtering out the incoming UV radiation from the Sun, can trap much of the infrared (IR) heat trying to escape the Earth to space. If stratospheric ozone is destroyed, ozone's contribution to the greenhouse effect is reduced.

This could offset some of the global warming due to man-made emissions of carbon dioxide, methane and nitrous oxide. Ironically, as the ozone layer gradually repairs itself during the 21st century, this cooling potential will be lost. More significantly, the replacement chemicals to CFCs, the HCFCs, which themselves do little harm to the ozone layer, are very strong greenhouse gases, and are further contributing to the potential problem of global warming.

HUMAN HEALTH

Ozone's unique physical properties allow the ozone layer to act as our planet's sunscreen, providing an invisible filter to help protect all life forms from the Sun's damaging ultraviolet (UV) rays. Most incoming UV radiation is absorbed by ozone and prevented from reaching the Earth's surface. Without the protective effect of ozone, life on Earth would not have evolved the way it has.

The ozone layer protects us from the harmful effects of certain wavelengths of ultraviolet (UV) radiation from the Sun. The danger to humans from UV radiation comes mainly from the UV-B range of the spectrum, although UV-A poses some risk if exposure is long enough. UV radiation is harmful to the eyes, can damage the immune system and over time can lead to the development of skin cancers. If ozone in the stratosphere is destroyed, more UV radiation will reach the Earth's surface, and incidences of these health effects will increase.

IMMUNE SYSTEM

UV radiation from the Sun can benefit health, generating vitamin D production in the skin. The required amount of radiation is, however, quite small. In summer, an exposure of 15 minutes to the hands and face is adequate. Vitamin D is also found in food. A normal diet will provide enough vitamin D for people even in winter.

In the treatment of some skin diseases such as psoriasis, UV radiation is being effectively exploited. Under a doctor's control, the benefit from the treatment is much greater than any consequential increase in skin cancer risk.

However, over exposure to UV radiation can impair the body's ability to fight off disease, in addition to causing cancer and a range of eye disorders. UV suppresses the immune system, irrespective of skin colour, making it easier for tumours to take hold and spread.

UV radiation suppresses allergic reactions of the skin and affects the immune system. When skin has been over-exposed to UV radiation, the activity of antibody-producing white blood cells is suppressed. These effects are not restricted to the part of skin actually subject to exposure, but may also occur on shielded parts of skin and throughout the whole immune system. As a result, the body fails to produce the antigens required for defence against a variety of diseases. This could have serious consequences, including a much-diminished effectiveness of vaccinations.

At the present time, the significance of a weakening of the immune system caused by UV radiation is not properly understood. The weakening can possibly act to promote the development of skin cancers and worsen infectious diseases stemming from bacteria, viruses and tropical parasites.

It may also activate viruses already present on the skin, such as herpes, and lead to an increase in diseases like measles, malaria, tuberculosis, leprosy and fungal infections, all of which have a stage involving the skin. People carrying the herpes virus should protect their faces against strong sunlight.

Scientific research suggests that sunburn can alter the distribution and function of disease-fighting white blood cells in humans for up to 24 hours after exposure to the Sun. In addition, repeated exposure to UV radiation may cause more long-lasting damage to the body's immune system.

Whilst little research has been conducted on the effects of decreasing stratospheric ozone on human immunity, it is likely that continued destruction of the ozone layer will lead to further health complications, in addition to skin cancers and eye disorders, as a result of the suppression of our ability to fight off disease.

Impacts of Ozone Depletion

Ultraviolet (UV) radiation from the Sun can cause a variety of health problems in humans, including skin cancers, eye cataracts and a reduction in the ability to fight off disease. Furthermore, UV radiation can be damaging to microscopic life in the surface oceans which forms the basis of the world's food chain, certain varieties of vegetation including rice and soya crops, and polymers used in paints, clothing and other materials.

A loss of ozone in the stratosphere because of mankind's pollution with ozone depleting chemicals such as CFCs will increase the amount of UV radiation that reaches the Earth's surface. As a consequence, health disorders, damage to plant and aquatic life, and degradation of materials will probably increase. Ozone depletion may even affect the global climate.

Land Plants

Excessive UV radiation inhibits the growth processes of almost all green plants. There is concern that ozone depletion may lead to a loss of plant species and reduce global food supply. Plants form the basis of the terrestrial food web, prevent soil erosion and water loss, and are the primary producers of oxygen and a primary removal sink for carbon dioxide, a greenhouse gas.

Exposure to UV radiation may have a dramatic effect on terrestrial plant life, although the impacts are at present poorly understood. Absorption of UV radiation varies widely from one organism to the next. In general UV radiation affects plant growth by reducing leaf size and limiting the area available for energy capture during photosynthesis. Plant stunting and a reduction in total dry weight are also typically seen in UV-irradiated plants, with a reduction in the nutrient content and the growth of the plants, especially in the pea and cabbage families.

A reduction in quality of certain types of tomato, potato, sugar beet and soya bean has also been observed. Forests also appear to be vulnerable. About half of the species of conifer seedlings so far studied have been adversely affected by UV radiation. Although old needles are able to protect themselves by strengthening their outer wax coating and by increasing the amount of protective pigment, young growing pine needles in contrast, suffer easily.

Indirect changes caused by UV radiation, such as flowering and germination rates, changes in plant form and how nutrients are distributed within the plant, may be more important than the damaging effects of the radiation itself.

Reliable scientific information on the effects of UV radiation on plants however, is limited. Only 4 out of 10 terrestrial plant ecosystems (temperate forest, agriculture, temperate grassland, and tundra/alpine ecosystems) have been studied. In addition, much of the existing information comes from greenhouses where plants are more sensitive to UV radiation than those grown outdoors.

There are indications that some weeds are more UV-resistant than crops. Many organisms have developed mechanisms for protecting themselves against over-exposure to UV radiation, for example by shielding themselves with pigment and repairing damaged DNA or plant tissue. However, for many organisms these mechanisms may not be sufficient to protect against increased levels of UV radiation as a result of ozone depletion.

Exposure tests made in USA and Australia have showed that over one hundred species of land plant could be sensitive to increases in UV radiation as a result of stratospheric ozone depletion. Some research has suggested that a 25% ozone depletion could result in a comparable reduction in total soya bean crop yield. International research has revealed that some species of rice suffer from even minor increases in UV radiation. Research into the efficient breeding and cultivation of strong species may help to offset some of the damaging effects of stratospheric ozone loss.

LEGISLATION

Under the terms of the Montreal Protocol developed nations have ceased production of new CFCs, halons and other ozone depleting chemicals (ODCs) to protect the ozone layer. Trade controls on the supply of these substances have been put in place to ensure compliance with the Protocol. Existing CFCs are re-used and recycled where possible. Nevertheless, the increasing price of CFCs as a result of the ban on new production has led to a wave of international smuggling.

Usually, when ozone-depleting substances are discarded or removed from equipment during the course of maintenance they become controlled waste. In Britain, the 1990 Environmental Protection Act ensured that waste chemicals which may contribute to stratospheric ozone depletion are disposed of as carefully as possibly to avoid any release to the atmosphere.

The production and consumption of new halons (ODCs containing bromine) has already ceased under the terms of the Montreal Protocol. However, whilst replacements have been developed these cannot be used in existing systems, which can only be maintained with recycled halons using surplus material from redundant installations. In the UK the Halon Users' National Consortium (HUNC) is managing the installed banks of halons, acting as a clearing house putting those who need to continue to use halons in contact with those who do not.

The Montreal Protocol and subsequent London and Copenhagen Amendments have demanded that existing CFCs should be recovered, recycled and re-used where possible. In the UK commercial users of refrigeration and air conditioning appliances can contact the Refrigeration Industry Board to ensure that best industrial practice is maintained during the disposal or re-use of CFCs.

Domestic users of old refrigerators can contact their local authority to find out if it operates a CFC recovery and recycling scheme.

The Montreal Protocol works through a system of trade barriers controlling supply to the market of ozone depleting chemicals. Imports of newly produced CFCs and halons by developed countries have already been banned, as have imports and exports in the ODCs carbon tetrachloride and 1,1,1 trichloroethane. Developing countries have been granted a period of grace to comply with the Montreal Protocol, to avoid undue stresses on their growing economies.

As a result of the decline in the production and use of CFCs, and the continuation of CFC production in developing countries (allowed under the provisions of the Montreal Protocol until 2010), the lure of illegal trade in CFCs is obvious. Significant volumes of illegal imports of CFCs into Western Europe have been reported, even though production in Western Europe ceased at the end of 1994.

Unfortunately, the Montreal Protocol currently does not require Parties (countries) to implement controls against illegal trade, although they have been urged to install verification programs to reduce illegal trade in ODCs.

MATERIALS DAMAGE

Ozone depletion will cause many materials to degrade faster. These materials include PVC (used in window and doorframes, pipes and gutters), nylon and polyester. They are all composed of compounds known as polymers. Synthetic polymers, naturally occurring biopolymers, as well as some other materials of commercial interest are adversely affected by UV radiation from the Sun.

Today's materials are somewhat protected from UV radiation by special additives. Therefore, any increase in UV levels as a result of ozone depletion will accelerate their breakdown, limiting their useful outdoor lifetime. UV radiation is mainly responsible for photo-damage ranging from discoloration to loss of mechanical integrity in polymers exposed to sunlight.

The use of higher levels of conventional light stabilisers in polymer-based materials is likely to be employed to minimise the effects of increased UV levels reaching the Earth's surface. However, it is not certain how resistant such light stabilisers are themselves to increased levels of UV radiation. In addition, their use will add to the cost of plastic products in target applications. With plastics rapidly displacing conventional materials in numerous applications, this is an important consideration particularly in the developing world.

It is not certain yet how other materials, including rubber, paints, wood, paper and textiles will be affected by increased UV radiation resulting from ozone depletion.

MEASURING OZONE DEPLETION

The most common stratospheric ozone measurement unit is the Dobson Unit (DU). The Dobson Unit is named after the atmospheric ozone pioneer G. M. B. Dobson who carried out the earliest studies on ozone in the atmosphere from the 1920s to the 1970s. A Dobson Unit measures the total amount of ozone in an overhead column of the atmosphere.

Dobson Units are measured by how thick the layer of ozone would be if it were compressed into one layer at 0 degrees Celsius and with a pressure of one atmosphere above it. Every 0. 01 millimetre thickness of the layer is equal to one Dobson Unit.

The average amount of ozone in the stratosphere across the globe is about 300 DU (or a thickness of only 3mm at 0°C and 1 atmospheric pressure!). Highest levels of ozone are usually found in the mid to high latitudes, in Canada and Siberia (360DU). When stratospheric ozone falls below 200 DU this is considered low enough to represent the beginnings of an ozone hole. Ozone holes of course commonly form during springtime above Antarctica, and to a lesser extent the Arctic.

MONITORING OZONE DEPLETION

Monitoring of the ozone layer has increased significantly since the 1980s when the Antarctic ozone hole was first discovered by the British Antarctic Survey. The ozone layer is monitored both by satellites and ground-based resources that are dedicated to observing the destruction of stratospheric ozone.

The main satellite that monitors the ozone layer is the TOMS (Total Ozone Mapping Spectrometer) satellite. The TOMS satellite measures the ozone levels from the back-scattered sunlight in the ultraviolet (UV) range. Another satellite is NASA's UARS (Upper Atmosphere Research Satellite) which was launched in September 1991.

This satellite is unique because it was configured to not only measure ozone levels, but also levels of ozone-depleting chemicals. GOME, launched in April 1995 on the ERS-2 satellite, marked the beginning of a long-term European ozone monitoring effort. Scientists receive high quality data on the global distribution of ozone and several other climate-influencing trace gases in the Earth's atmosphere.

In 1987, Canada became the first country in the world to focus on the Arctic ozone layer, following the discovery of the ozone hole over the Antarctic. A cross-country network of monitoring stations has kept continuous watch on Canada's ozone layer for more than three decades. The existence of these early records, before any major human influence on the upper atmosphere, is vital to understanding the changes that have occurred in the ozone layer.

In the UK, stratospheric ozone levels are monitored every winter and spring at Cambourne in Cornwall and Lerwick in the Shetland Isles.

MONTREAL PROTOCOL

In 1985 the Vienna Convention established mechanisms for international co-operation in research into the ozone layer and the effects of ozone depleting chemicals (ODCs). 1985 also marked the first discovery of the Antarctic ozone hole. On the basis of the Vienna Convention, *the Montreal Protocol on Substances that Deplete the Ozone Layer* was negotiated and signed by 24 countries and by the European Economic Community in September 1987. The

Protocol called for the Parties to phase down the use of CFCs, halons and other man-made ODCs.

The Montreal Protocol represented a landmark in the international environmentalist movement. For the first time whole countries were legally bound to reducing and eventually phasing out altogether the use of CFCs and other ODCs. Failure to comply was accompanied by stiff penalties. The original Protocol aimed to decrease the use of chemical compounds destructive to ozone in the stratosphere by 50% by the year 1999.

The Protocol was supplemented by agreements made in London in 1990 and in Copenhagen in 1992, where the same countries promised to stop using CFCs and most of the other chemical compounds destructive to ozone by the end of 1995. Fortunately, it has been fairly easy to develop and introduce compounds and methods to replace CFC compounds.

In order to deal with the special difficulties experienced by developing countries it was agreed that they would be given an extended period of grace, so long as their use of CFCs did not grow significantly.

China and India, for example, are strongly increasing the use of air conditioning and cooling devices. Using CFC compounds in these devices would be cheaper than using replacement compounds harmless to ozone. An international fund was therefore established to help these countries introduce new and more environmentally friendly technologies and chemicals.

The depletion of the ozone layer is a worldwide problem which does not respect the frontiers between different countries. It can only be affected through determined international co-operation.

The Timetable

Montreal Protocol (1987)

CFCs (11, 12, 113, 114, 115): Phase down 1986 levels by 20% by 1994; 50% by 1999.

London Amendment (1990)

CFCs 13, 111, 112, 211, 212, 213, 214, 215, 216, 217: Phase down 1989 levels 20% by 1993; 85% by 1997; 100% by 2000.

Halons (1211, 1301, 2402): Phase down 1986 levels 50% by 1995; 100% by 2000.

Carbon Tetrachloride: Phase down 1989 levels 85% by 1995; 100% by 2000.

Copenhagen Amendment (1992)

CFCs: phase out by 1995

Halons: phase out by 1993

Carbon Tetrachloride: phase out by 1995

HCFCs: phase down 1989 levels 35% by 2004; 90% by 2019; 100% by 2029.

The Montreal Protocol has been further adjusted in Vienna (1995), Montreal (1997) and most recently in Beijing (1999). The Beijing Amendment (1999) has introduced a freezing of HCFC production by 2003.

OZONE DEPLETING CHEMICALS

The most commonly known ozone depletion chemicals (ODCs) are the CFCs or chlorofluorocarbons. Over the last 30 years man-made CFCs have been the main cause of stratospheric ozone depletion. Fortunately, CFCs, once used for example in refrigeration, air conditioning, and fire extinguishers, have been largely phased out in the aftermath of the Montreal Protocol.

CFCs however, are not the only ozone-depleting chemicals. Other ODCs include the methylhalides, carbon tetrachloride (CCl_4), carbon tetrafluoride (CF_4), and the halons which contain bromine instead of chlorine. Such compounds are called halocarbons.

Carbon tetrachloride (CCl_4), despite its toxicity, was first used in the early 1900s as a fire extinguishant, and more recently as an industrial solvent, an agricultural fumigant, and in many other industrial processes including petrochemical refining, and pesticide and pharmaceuticals production. Recently it has also been used in the production of CFC-11 and CFC-12.

It has accounted for less than 8% of total ozone depletion. The use of carbon tetrachloride in developed countries however, has been prohibited since the beginning of 1996 under the Montreal Protocol.

Methyl chloroform, also known as 1 1,1 trichloroethane is a versatile, all-purpose industrial solvent used primarily to clean metal and electronic parts. It was introduced in the 1950s as a substitute for carbon tetrachloride. Methyl chloroform has accounted for roughly 5% of total ozone depletion. The use of methyl chloroform in developed countries has been prohibited since the beginning of 1996 under the Montreal Protocol.

Halons, unlike CFCs, contain bromine, which also destroys ozone in the stratosphere. Halons are used primarily in fire extinguishers. Halon-1301 has an ozone depleting potential 10 times that of CFC-11. Although the use of halons in developed countries has been phased out since 1996, the atmospheric concentration of these potent, ozone destroyers is still rising because of their long atmospheric lifetimes.

To date halons have accounted for about 5% of global ozone depletion. Methyl bromide, another bromine-containing halocarbon, has been used as a pesticide since the 1960s. Today, scientists estimate that human sources of methyl bromide have been responsible for approximately 5 to 10% of global ozone depletion.

OZONE

Without ozone, life on Earth would not have evolved in the way it has. The first stage of single cell organism development requires an oxygen-free environment. This type of environment existed on earth over 3000 million years ago. As the primitive forms of plant life multiplied and evolved, they began to release minute amounts of oxygen through the photosynthesis reaction (which converts carbon dioxide into oxygen).

The build up of oxygen in the atmosphere led to the formation of the ozone layer in the upper atmosphere or stratosphere. This layer filters out incoming radiation in the "cell-damaging" ultraviolet (UV) part of the spectrum. Thus with the development of the ozone layer came the formation of more advanced life forms.

Ozone is a form of oxygen. The oxygen we breathe is in the form of oxygen molecules (O_2) - two atoms of oxygen bound together. Normal oxygen which we breathe is colourless and odourless. Ozone, on the other hand, consists of three atoms of oxygen bound together (O_3). Most of the atmosphere's ozone occurs in the region called the stratosphere.

Ozone is colourless and has a very harsh odour. Ozone is much less common than normal oxygen. Out of 10 million air molecules, about 2 million are normal oxygen, but only 3 are ozone.

Most ozone is produced naturally in the upper atmosphere or stratosphere. While ozone can be found through the entire atmosphere, the greatest concentration occurs at altitudes between 19 and 30 km above the Earth's surface. This band of ozone-rich air is known as the "ozone layer".

Ozone also occurs in very small amounts in the lowest few kilometres of the atmosphere, a region known as the troposphere. It is produced at ground level through a reaction between sunlight and volatile organic compounds (VOCs) and nitrogen oxides (NOx), some of which are produced by human activities such as driving cars. Ground-level ozone is a component of urban smog and can be harmful to human health.

Even though both types of ozone contain the same molecules, their presence in different parts of the atmosphere has very different consequences. Stratospheric ozone blocks harmful solar radiation - all life on Earth has adapted to this filtered solar radiation. Ground-level ozone, in contrast, is simply a pollutant. It will absorb some incoming solar radiation, but it cannot make up for ozone losses in the stratosphere.

OZONE HOLE

In some of the popular news media, as well as in many books, the term "ozone hole" has and often still is used far too loosely. Frequently, the term is employed to describe any episode of ozone depletion, no matter how minor.

Unfortunately, this sloppy language trivialises the problem and blurs the important scientific distinction between the massive ozone losses in polar regions and the much smaller, but nonetheless significant, ozone losses in other parts of the world.

Technically, the term "ozone hole" should be applied to regions where stratospheric ozone depletion is so severe that levels fall below 200 Dobson Units (D. U.), the traditional measure of stratospheric ozone.

Normal ozone concentration is about 300 to 350 D. U. Such ozone loss now occurs every springtime above Antarctica, and to a lesser extent the Arctic, where special meteorological conditions and very low air temperatures accelerate and enhance the destruction of ozone loss by man-made ozone depleting chemicals (ODCs).

OZONE LAYER

The ozone layer is not really a layer at all, but has become known as such because most ozone particles are scattered between 19 and 30 kilometres (12 to 30 miles) up in the Earth's atmosphere, in a region called the stratosphere.

The concentration of ozone in the ozone layer is usually under 10 parts ozone per million. Without the ozone layer, a lot of ultraviolet (UV) radiation from the Sun would not be stopped reaching the Earth's surface, causing untold damage to most living species. In the 1970s, scientists discovered that chlorofluorocarbons (CFCs) could destroy ozone in the stratosphere.

Ozone is created in the stratosphere when UV radiation from the Sun strikes molecules of oxygen (O_2) and causes the two oxygen atoms to split apart. If a freed atom bumps into another O_2, it joins up, forming ozone (O_3). This process is known as photolysis. Ozone is also naturally broken down in the stratosphere by sunlight and by a chemical reaction with various compounds containing nitrogen, hydrogen and chlorine.

These chemicals all occur naturally in the atmosphere in very small amounts.

In an unpolluted atmosphere there is a balance between the amount of ozone being produced and the amount of ozone being destroyed. As a result, the total concentration of ozone in the stratosphere remains relatively constant. At different temperatures and pressures (i. e. varying altitudes within the stratosphere), there are different formation and destruction rates. Thus, the amount of ozone within the stratosphere varies according to altitude. Ozone concentrations are highest between 19 and 23 km.

Most of the ozone in the stratosphere is formed over the equator where the level of sunshine striking the Earth is greatest. It is transported by winds towards higher latitudes. Consequently, the amount of stratospheric ozone above a location on the Earth varies naturally with latitude, season, and from day-to-day.

Under normal circumstances highest ozone values are found over the Canadian Arctic and Siberia, whilst the lowest values are found around the equator. The ozone layer over Canada is normally thicker in winter and early spring, varying naturally by about 25% between January and July. Weather conditions can also cause considerable daily variations.

POLAR STRATOSPHERIC CLOUDS

Polar stratospheric clouds (PSCs) are important components of the ozone depletion process in the polar regions of Antarctica and the Arctic. PSCs provide the surfaces upon which chemical reactions involved in ozone destruction take place. These reactions lead to the production of free chlorine and bromine, released from CFCs and other ozone depleting chemicals (ODCs), which directly destroy ozone molecules.

Polar stratospheric clouds (PSCs), also known as mother-of-pearl or nacreous clouds, have been observed for many years. As their name suggests, the clouds form in cold polar stratospheric winters where, despite the dryness of the stratosphere, the temperature drops low enough for condensation to occur. PSCs are believed to be made of nitric acid and ice. In other parts of the world the stratosphere is too warm for these clouds to form, which is one reason why the "ozone hole" is confined to the Antarctic region.

POLAR VORTEX

Polar regions experience the greatest changes in ozone concentrations, with severe ozone depletion occurring each springtime, particularly over Antarctica. This is due significantly to the special meteorological conditions which exist in these regions of the world.

The topography and circular shape of Antarctica is such that a stagnant whirlpool of extremely cold stratospheric air forms over the region during the long polar night. The air circulates within this polar vortex all winter, becoming cold enough to allow the formation of polar stratospheric clouds which accelerate and enhance stratospheric ozone destruction when sunlight returns during early spring.

Such a vortex also exists in the Arctic region, but to a lesser degree. The Arctic region consists of many distinct landmasses and islands spread out around the North Pole, and so the air cannot circulate quite as easily.

PROTECTION AGAINST THE SUN

- There are a number of things that we can do to safeguard our health against the effects of increasing UV radiation from the Sun.
- Sunglasses that provide 99-100% UV-A and UV-B protection will greatly reduce sunlight exposure that can lead to cataracts and other eye damage. Check the label when buying sunglasses.
- A hat with a wide brim offers good sun protection to your eyes, ears, face, and the back of your neck - areas particularly prone to overexposure.
- Tightly woven, loose-fitting clothes offer excellent protection against UV radiation. Any clothing is better than none at all.
- A sunscreen with protection factor of at least 15 blocks most harmful UV radiation. Apply sunscreen liberally and reapply every 2 hours when working, playing, or exercising outdoors. Even waterproof sunscreen can come off when you dry yourself off with a towel.
- The Sun's UV rays are strongest between 10 a. m. and 4 p. m. To the extent you can, limit exposure to the Sun during these hours.
- Sunlamps damage the skin and unprotected eyes and are best avoided entirely.

SEA LIFE

Plankton form the foundation of aquatic food webs. Plankton are generally found in the upper layer of the oceans in which there is sufficient sunlight to support the photosynthesis of food. Since UV radiation has the ability to penetrate up to 20 metres down in clear water, plankton and other light dependent organisms often experience cell damage, much as human DNA can be damaged by the strong solar radiation.

Both plant (phytoplankton) and animal (zooplankton) species are damaged by UV radiation even at current levels. Since UV radiation is absorbed by only a few layers of cells, large organisms are more protected, whilst smaller ones, such as plankton are among the most severely affected by UV radiation. As plankton make up the base of the marine food chain, changes in their number and species composition will influence fish and shellfish production worldwide. These kinds of losses will have a direct impact on the food supply.

UV radiation has also been found to cause damage to the early developmental stages of fish, shrimp, crab, amphibians and other animals. The most severe effects are decreased reproductive capacity and impaired larval development. Even at current levels, UV radiation is a limiting factor, and small increases in UV exposure could result in a significant reduction in the size of the population of animals that eat these smaller creatures.

Research indicates that many plankton species already seem to be at or near their maximum tolerance of UV radiation. Thus, even small increases in UV levels as a result of ozone depletion may have a dramatic impact on plankton life and on entire marine ecosystems. If ozone layer depletion reached 15% over temperate waters in the mid-latitudes, it would take fewer than five days in summer for half the zooplankton in the top metre of these waters to die from the increased radiation.

Additionally, large amounts of young fish, shrimp and crabs would die before reaching their reproductive age. Less food would be available for adult fish and other higher forms of marine life, and therefore for human consumption. This is of particular relevance, as more than 30% of the world's animal protein for human consumption comes from the sea.

Effects of the ozone hole in Antarctica have already been seen in some of the organisms. Most of the Antarctic organisms have a low tolerance for UV radiation since for much of the year, hardly any direct sunlight reaches the continent. With the reduced ozone in springtime, UV radiation has been able to penetrate the atmosphere with a higher intensity. Already at the base of the Antarctica food chain an impact has been felt. Increased UV radiation has already reduced the plankton populations by between 6% and 12%. Consequently, species higher up have felt the impact.

SKIN CANCER

The most well known effect of UV radiation is the slight reddening or burning of the skin in sunshine. This change of colour is caused by an expansion of the skin's blood vessels. For most people burning is followed by tanning within a couple of days. A permanent tan will occur when the UV radiation causes a pigment called melanin to form in the pigment cells of the skin.

Over a period of years, exposure to radiation originating from the Sun causes damage in the skin's connective tissues, so-called photo-aging. This shows itself as a thickening of the skin, as wrinkles and decreasing elasticity. UV radiation increases the risk of getting skin cancer.

Research has shown that even small amounts of UV radiation can cause considerable harm. UV damages the genetic material of DNA and is related to some types of skin cancer. It is important to note however, that UV radiation has always had this effect on humans. In recent years non-melanoma skin cancer has become more prevalent in many parts of the world because people are spending more time in the Sun and are exposing more of their skin in the process.

The relationship between the occurrence of milder non-melanoma skin cancers and time spent in the Sun is well documented. Such cancers generally occur in people in their 70s and 80s on areas of the skin usually exposed to sunlight (such as the face or hands). Malignant melanoma however, usually occurs in younger people and in skin areas not necessarily exposed to sunlight.

It tends to occur most commonly among groups of people less likely to have spent significant amounts of time outdoors.

The risk of developing malignant melanoma is directly related to the sensitivity of an individual's skin to the Sun. Fair-skinned individuals are more susceptible than darker-skinned individuals. The victims are almost exclusively Caucasians, particularly fair-skinned Caucasians. The incidence of malignant melanoma has been increasing among light-skinned populations around the world for decades.

Ozone in the stratosphere protects Earth from damaging amounts of UV radiation. A depleted ozone layer would allow more of the Sun's rays to reach Earth's surface. An increase in the levels of UV reaching the Earth as a result of ozone depletion may compound the effects of spending more time in the Sun. According to some estimates a sustained 10% global loss of ozone may lead to a 26% increase in the incidence of skin cancers among fair skinned people.

Australia, with high sunshine levels, has very high skin cancer rates. An estimated 2 out of every 3 people in most parts of the country will develop some form of skin cancer. In Queensland, where sunshine levels are greatest, the probability jumps to 3 in every 4. In 1930s America the chances of developing the more serious malignant melanoma was 1 in 1500. In 1991 it had soared to 1 in 150, and it is predicted to rise to as much as 1 in 75 early in the 21st century.

STRATOSPHERE

The stratosphere is the second major layer of the atmosphere and lies above the troposphere, the lowest layer. It occupies the region of atmosphere from about 12 to 50 km above the Earth's surface, although its lower boundary tends to be higher nearer the equator and lower nearer the poles.

Most ozone in the atmosphere is found in the stratosphere in a region between 19 and 30 km above the Earth's surface, known as the ozone layer. Stratospheric ozone absorbs ultraviolet (UV) radiation from the Sun providing an invisible filter to help protect all life forms from the Sun's damaging rays.

SUN

The Sun is the source of ultraviolet (UV) radiation which can be so damaging to life on Earth. Fortunately, the ozone layer protects us and other life forms against the damaging affects of solar rays.

Changes in solar activity can influence naturally the amount of ozone in the stratosphere. Stratospheric ozone is primarily created by ultraviolet (UV) radiation coming from the Sun. The Sun's energy release does vary, especially over the 11-year sunspot cycle. During the active phase of the 11-year sunspot cycle, more ozone is produced in the ozone layer. This phenomenon can boost the average ozone concentration over the poles by about 4%, although the global increase is nearer 2%.

Unusual solar activity can cause the ozone levels in the upper stratosphere to be substantially depleted, but since most of the ozone is in the middle stratosphere, the effect on total stratospheric ozone is negligible.

ULTRAVIOLET RADIATION

Ultraviolet radiation is a form of electromagnetic energy. The various forms of energy, or radiation, are classified according to wavelength, measured in nanometres (nm). One nanometre is a millionth of a millimetre. The shorter the wavelength, the more energetic the radiation. In order of decreasing energy, the principal forms of radiation are gamma rays, X rays, ultraviolet (UV) radiation, visible light, infrared (IR) radiation, microwaves and radio waves. Most energy from the Sun is in the form of visible and UV radiation.

There are three categories of UV radiation:

- UV-A, between 320 and 400 nm
- UV-B, between 280 and 320 nm
- UV-C, between 200 and 280 nm

Generally, the shorter the wavelength, the more biologically damaging UV radiation can be if it reaches the Earth in sufficient quantities. UV-A is the least damaging (longest wavelength) form of UV radiation and reaches the Earth in greatest quantity. Most UV-A rays pass right through the ozone layer in the stratosphere. UV-B radiation can be very harmful. Fortunately, most of the Sun's UV-B radiation is absorbed by ozone in the stratosphere. UV-C radiation is potentially the most damaging because it is very energetic. Fortunately, all UV-C is absorbed by oxygen and ozone in the stratosphere and never reaches the Earth's surface.

Although the ozone layer is the one constant defence against UV penetration, several other factors can effect the amount of UV radiation arriving at the Earth's surface. The Sun's rays are the most intense near the equator where they impact the Earth's surface at the most direct angle. At higher latitudes, the Sun is lower in the sky during daylight hours and the strength of radiation is weaker. Furthermore, during the winter months, the Sun's rays strike at a more oblique angle than they do in the summer. This means that all solar radiation travels a longer path through the atmosphere to reach the Earth, and is therefore less intense. Daily changes in the angle of the Sun influence the amount of UV radiation that passes through the atmosphere. When the Sun is low in the sky, its rays must travel a greater distance through the atmosphere and may be scattered and absorbed by water vapour and other atmospheric components. The greatest amount of UV reaches the Earth around midday when the Sun is at its highest point.

Other influences on UV radiation include altitude, cloud cover, air pollution and land type. On the top of a mountain the air is thinner and cleaner. Consequently more UV reaches there than at lower elevations. Clouds can have a marked impact on the amount of UV radiation that reaches the Earth's surface. Thick clouds block more UV than thin cloud cover.

Like clouds, urban smog can reduce the amount of UV radiation reaching the Earth. Finally, incoming UV radiation is reflected from most surfaces. Snow reflects up to 85%, whilst other land types reflect much less. Water reflects only 5%. Reflected UV can damage people, plants, and animals just as direct UV does.

VOLCANOES

Volcanic eruptions are one of the few natural things that can have a diminishing effect on the ozone layer. Large eruptions can potentially inject significant quantities of chlorine (via hydrochloric acid - HCl) directly in the stratosphere where the highest concentrations of ozone are found. However, the vast majority of volcanic eruptions are too weak to reach the stratosphere, beginning approximately 12 km above the Earth's surface.

Thus, any HCl emitted during a smaller eruption remains in the lowest atmosphere where it is quickly dissolved and washed out by rain. In contrast man-made CFCs do not dissolve in water and can therefore reach the stratosphere through atmospheric mixing.

It is also possible that ice particles containing sulphuric acid from large volcanic eruptions may contribute to ozone loss. When chlorine compounds resulting from the breakup of man-made CFCs in the stratosphere are present, the sulphate particles serve to convert them into more active forms that may cause more rapid ozone depletion. In 1991 Mt. Pinatubo in the Philippines erupted tonnes of dust and gas high into the atmosphere which caused global reductions in the ozone layer for 2 to 3 years.

Thus, whilst large volcanic eruptions may increase the rate of stratospheric ozone depletion, it is more probable that the presence of chlorine from man-made CFC emissions is the chief cause of ozone loss in the first instance.

Chapter 11

Deforestation

Deforestation is the removal or destruction of large areas of forest or rainforest. Deforestation happens for many reasons, such as logging, agriculture, natural disasters, urbanization and mining. There are several ways to clear forest — burning and clear-cutting the land are two methods.

Although deforestation occurs worldwide, it's a particularly critical issue in the Amazon rainforests of Brazil. There, the tropical forests, and the species of plants and animals within them, are disappearing at an alarming rate. In December 2007, for example, experts measured Amazon destruction at more than 360 square miles (932 square kilometers) in just one month.

The effects of deforestation are long lasting and devastating. Entire species of insects and animals have disappeared because of the destruction of their habitats. Deforestation can cause catastrophic flooding as well. And scientists see that deforestation has a significant effect on climate change, or global warming.

If deforestation is so destructive, then why is it even done? What's driving the destruction of forests? To learn more about the causes and effects of deforestation, as well as current conservation initiatives and solutions, keep reading.

There are several reasons deforestation occurs: trees or derived charcoal can be sold as a commodity and used by humans, while cleared land is used as pasture, plantations of commodities and human settlement. The removal of trees without sufficient reforestation has resulted in damage to habitat, biodiversity loss and aridity. Deforested regions often degrade into wasteland.

Disregard or ignorance of intrinsic value, lack of ascribed value, lax forest management and deficient environmental law are some of the factors that allow deforestation to occur on a large scale. In many countries, deforestation is an ongoing issue that is causing extinction, changes to climatic conditions, desertification, and displacement of indigenous people.

Fig. A Deforested Site

However, among countries with a per capita GDP of at least $4,600, net deforestation rates have ceased to increase.

CAUSES OF DEFORESTATION

For the most part, human activity is to blame for deforestation, though natural disasters do play a role. So let's take a look at how and why humans deforest areas.

Logging, or cutting down trees in a forest to harvest timber for wood, products or fuel, is a primary driver of deforestation. Logging affects the environment in several ways. Since trucks and large equipment need to get into the forest in order to access trees and transport timber, loggers must clear large areas for roadways.

Selective logging — where only the most valuable trees are felled — doesn't help matters, as one falling tree can bring down dozens of surrounding trees and thin the forest's protective canopy. The forest canopy is important to the forest's ecosystem because it houses and protects plant, animal and insect populations. It also protects the forest floor, which slows down soil erosion.

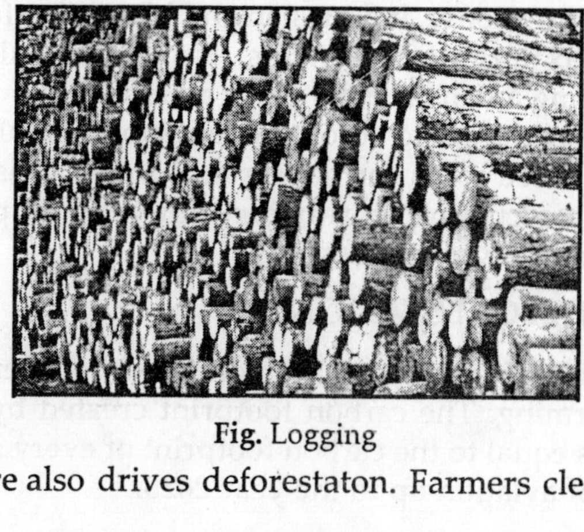

Fig. Logging

Agriculture also drives deforestaton. Farmers clear the land for crops or for cattle and often will clear acres of land using slash and burn techniques — cutting down trees and then burning them. Migratory farmers clear a forest area and use it until the soil becomes too degraded for crops. Then they move on and clear a new patch of forest. The abandoned land, if left untouched, will eventually reforest, but it will take many, many years to return to its original state.

Hydroelectric dams are quite controversial because while they help to power communities, they also contribute to deforestation. Damming opponents believe that the building of such structures not only has a negative environmental impact, but it also opens up the area to loggers and more roads. To build a hydroelectric dam, acres of land must be flooded, which causes decomposition and release of greenhouse gases. Local people can also be displaced by dam projects, causing further deforestation when these people resettle elsewhere.

Fires, both accidental and intended, destroy acres of forest very quickly. Areas affected by logging are more susceptible to fires due to the number of dried, dead trees. Milder winters and extended warm seasons due to global warming also fuel fires. For example, certain species of beetle that usually die off each winter are now able to survive and continue feeding on trees. This feeding causes the trees to die and dry out, making them into kindling.

Mining also results in deforestation. Digging a coal, diamond or gold mine requires the removal of all forest cover, not just for the mines but also for trucks and equipment. Recently, Venezuela denied a corporation called Crystallex permission to dig a mine because of environmental concerns.

Palm oil has been receiving attention lately for its potential as a biofuel and is used in many packaged foods and beauty products. But palm oil is another cause of deforestation. Its rising prices make it more valuable, and, in response, Indonesian and Malaysian farmers destroy acres of trees to harvest it. For this reason, several countries are currently debating a ban on palm oil as a biofuel.

As cities grow larger to accommodate more people, trees are cut down to make more room for houses and roads. This urban sprawl deforestation is occurring worldwide, now that 50 per cent of the world's population lives in cities

EFFECTS OF DEFORESTATION

Scientists are finding more and more links between deforestation and global warming. The carbon footprint created by four years of deforestation is equal to the carbon footprint of every single air flight in the history of aviation up to the year 2025.

Let's break that down into simple logic: Trees absorb carbon dioxide. So fewer trees means more carbon dioxide is loose in the air. More carbon dioxide means an increased greenhouse effect, which leads to global warming.

Fig. Deforestation Results in Soil Erosion

Reduced biodiversity is another deforestation concern. Rainforests, arguably the biggest victims of deforestation, cover only about 7 per cent of the world's surface. However, within this 7 per cent live almost half of all plant and animal species on earth. Some of these species only live in small specific areas, which makes them especially vulnerable to extinction.

As the landscape changes, some plants and animals are simply unable to survive. Species from the tiniest flower to large orangutans are becoming endangered or even extinct. Biologists believe that the key to curing many diseases resides within the biology of these rare plants and animals, and preservation is crucial.

Soil erosion, while a natural process, accelerates with deforestation. Trees and plants act as a natural barrier to slow water as it runs off the land. Roots bind the soil and prevent it from washing away. The absence of vegetation causes the topsoil to erode more quickly. It's difficult for plants to grow in the less nutritious soil that remains.

Because trees release water vapour into the atmosphere, fewer trees means less rain, which disrupts the water table (or groundwater level). A lowered water table can be devastating for farmers who can't keep crops alive in such dry soil.

On the other hand, deforestation can also cause flooding. Coastal vegetation lessens the impact of waves and winds associated with a storm surge. Without this vegetation, coastal villages are susceptible to damaging floods. The 2008 cyclone in Myanmar proved this fact to catastrophic effect. Scientists believe that the removal of coastal mangrove forests over the past decade caused the cyclone to hit with much more force.

Deforestation also affects indigenous people, both physically and culturally. Because many indigenous people actually have no legal rights to the land on which they live, governments that want to use the forest for profit can actually "evict" them. As these populations leave the rainforest, they also leave their culture behind

WAYS TO REDUCE DEFORESTATION AND REPAIR THE DAMAGE

In December 2007, the United Nations Climate Change Conference took place in Bali, Indonesia. After 10 days of intense discussion, more than 180 countries agreed to the Bali Roadmap. The Bali Roadmap will guide participating countries in emissions reduction and intends to lead to a binding agreement at the 2009 United Nations summit in Denmark. The United States and China initially did not agree to mandatory reductions, wanting countries to set their own goals, but they eventually conceded.

The roadmap includes specific measures to reduce deforestation — for tropical rainforests in particular. Many developing countries' economies rely on their forests, and they argue they should be able to use their land as they please. In response, the roadmap will investigate policies to financially reward countries who reduce their emissions by a certain per centage (the per centage has not yet been determined). Even this proposal faces controversy, however.

Because those countries with the highest baseline rate of deforestation will receive the most reward credits, critics fear that many countries will rush to cut down trees in order to raise their own baseline.

Besides the U. N. , there also are dozens of nonprofits working to combat deforestation.

A few well-known organizations include:

- *Conservation International*: Teaches local farmers how to maximize their existing land, rather than clear new areas
- *The World Wildlife Fund*: Works to shape policies and teams with communities to preserve forests
- *Rainforest Action Network*: Uses in-your-face advertising campaigns to call attention to the rainforests
- *The Environmental Defence Fund*: Champions government bills that provide financial incentive to private landowners (such as farmers) who practice land conservation
- *The Sierra Club*: Works to protect and restore U. S. forests

- *Amazon Watch*: Defends the rights of indigenous people and communities faced with industrial development
- *The Nature Conservancy*: Has developed several initiatives to advance conservation

Can we really save the forests? Once the trees are gone, is it possible to restore the land? Most deforested areas, if left alone, will eventually regenerate to fertile landscape. We can certainly plant more trees — a process called reforestation. In fact, many nonprofit organizations have popped up to support reforestation. For example, Carbonfund. org currently works on reforesting areas like Nicaragua and the state of Louisiana.

In the meantime, new movements in forest protection have sprung up over the years.

They include:

- *Eco-forestry*: Where only carefully selected trees are cut down and are transported with minimal damage to the area; the forest ecosystem is preserved while commercial timber extraction is still permitted
- *Green business*: Focuses on recycled paper and wood products, wood alternatives and environmentally responsible consumerism
- *Land use planning*: Advocates environmentally friendly development techniques, such as reduction of urban and suburban sprawl
- *Community forestry*: Where concerned citizens come together to manage and participate in keeping their local forests viable and sustainable

CAUSES OF ANTHROPOGENIC DEFORESTATION

There are many root causes of deforestation, including corruption of government institutions,the inequitable distribution of wealth and power,population growthand overpopulation,and urbani-zation. Globalization is often viewed as another root cause of deforestation,though there are cases in which the impacts of globalization (new ûows of labour, capital, commodities, and ideas) have promoted localized forest recovery.

In 2000 the United Nations Food and Agriculture Organization (FAO) found that "the role of population dynamics in a local setting may vary from decisive to negligible," and that deforestation can result from "a combination of population pressure and stagnating economic, social and technological conditions."

According to British environmentalist Norman Myers, 5% of

deforestation is due to cattle ranching, 19% due to over-heavy logging, 22% due to the growing sector of palm oil plantations, and 54% due to slash-and-burn farming.

The degradation of forest ecosystems has also been traced to economic incentives that make forest conversion appear more profitable than forest conservation. Many important forest functions have no markets, and hence, no economic value that is readily apparent to the forests' owners or the communities that rely on forests for their well-being. From the perspective of the developing world, the benefits of forest as carbon sinks or biodiversity reserves go primarily to richer developed nations and there is insufficient compensation for these services.

Developing countries feel that some countries in the developed world, such as the United States of America, cut down their forests centuries ago and benefited greatly from this deforestation, and that it is hypocritical to deny developing countries the same opportunities: that the poor shouldn't have to bear the cost of preservation when the rich created the problem.

Experts do not agree on whether industrial logging is an important contributor to global deforestation. Similarly, there is no consensus on whether poverty is important in deforestation. Some argue that poor people are more likely to clear forest because they have no alternatives, others that the poor lack the ability to pay for the materials and labour needed to clear forest. Claims that population growth drives deforestation have been disputed; one study found that population increases due to high fertility rates were a primary driver of tropical deforestation in only 8% of cases.

ENVIRONMENTAL PROBLEMS

Atmospheric

Deforestation is ongoing and is shaping climate and geography.

Deforestation is a contributor to global climate change, and is often cited as one of the major causes of the enhanced greenhouse effect. Tropical deforestation is responsible for approximately 20% of world greenhouse gas emissions.

According to the Intergovernmental Panel on Climate Change deforestation, mainly in tropical areas, account for up to one-third of total anthropogenic carbon dioxide emissions. Trees and other plants remove carbon (in the form of carbon dioxide) from the atmosphere during the process of photosynthesis and release it back into the atmosphere during normal respiration.

Only when actively growing can a tree or forest remove carbon over an annual or longer timeframe. Both the decay and burning of wood releases much of this stored carbon back to the atmosphere. In order for forests to take up carbon, the wood must be harvested and turned into long-lived products and trees must be re-planted. Deforestation may cause carbon stores held in soil to be released. Forests are stores of carbon and can be either sinks or sources depending upon environmental circumstances. Mature forests alternate between being net sinks and net sources of carbon dioxide.

Reducing emissions from the tropical deforestation and forest degradation (REDD) in developing countries has emerged as new potential to complement ongoing climate policies. The idea consists in providing financial compensations for the reduction of greenhouse gas (GHG) emissions from deforestation and forest degradation".

Rainforests are widely believed by laymen to contribute a significant amount of world's oxygen, although it is now accepted by scientists that rainforests contribute little net oxygen to the atmosphere and deforestation will have no effect on atmospheric oxygen levels. However, the incineration and burning of forest plants in order to clear land releases tonnes of CO_2, which contributes to global warming.

Forests are also able to extract carbon dioxide and pollutants from the air, thus contributing to biosphere stability.

HYDROLOGICAL

The water cycle is also affected by deforestation. Trees extract groundwater through their roots and release it into the atmosphere. When part of a forest is removed, the trees no longer evaporate away this water, resulting in a much drier climate. Deforestation reduces the content of water in the soil and groundwater as well as atmospheric moisture. Deforestation reduces soil cohesion, so that erosion, flooding and landslides ensue. Forests enhance the recharge of aquifers in some locales, however, forests are a major source of aquifer depletion on most locales.

Shrinking forest cover lessens the landscape's capacity to intercept, retain and transpire precipitation. Instead of trapping precipitation, which then percolates to groundwater systems, deforested areas become sources of surface water runoff, which moves much faster than subsurface flows. That quicker transport of surface water can translate into flash flooding and more localized floods than would occur with the forest cover.

Deforestation also contributes to decreased evapotranspiration, which lessens atmospheric moisture which in some cases affects precipitation levels down wind from the deforested area, as water is not recycled to downwind forests, but is lost in runoff and returns directly to the oceans. According to one preliminary study, in deforested north and northwest China, the average annual precipitation decreased by one third between the 1950s and the 1980s.

Trees, and plants in general, affect the water cycle significantly:

- Their canopies intercept a proportion of precipitation, which is then evaporated back to the atmosphere (canopy interception);
- Their litter, stems and trunks slow down surface runoff;
- Their roots create macropores - large conduits - in the soil that increase infiltration of water;
- They contribute to terrestrial evaporation and reduce soil moisture via transpiration;
- Their litter and other organic residue change soil properties that affect the capacity of soil to store water.
- Their leaves control the humidity of the atmosphere by transpiration. 99% of the water pulled up by the roots move up to the leaves for transpiration.

As a result, the presence or absence of trees can change the quantity of water on the surface, in the soil or groundwater, or in the atmosphere. This in turn changes erosion rates and the availability of water for either ecosystem functions or human services.

The forest may have little impact on flooding in the case of large rainfall events, which overwhelm the storage capacity of forest soil if the soils are at or close to saturation. Tropical rainforests produce about 30% of our planets fresh water.

SOIL

Undisturbed forest has very low rates of soil loss, approximately 2 metric tons per square kilometre (6 short tons per square mile). Deforestation generally increases rates of soil erosion, by increasing the amount of runoff and reducing the protection of the soil from tree litter. This can be an advantage in excessively leached tropical rain forest soils. Forestry operations themselves also increase erosion through the development of roads and the use of mechanized equipment.

China's Loess Plateau was cleared of forest millennia ago. Since then it has been eroding, creating dramatic incised valleys, and providing the sediment that gives the Yellow River its yellow colour and that causes the flooding of the river in the lower reaches (hence the river's nickname 'China's sorrow').

Removal of trees does not always increase erosion rates. In certain regions of southwest US, shrubs and trees have been encroaching on grassland. The trees themselves enhance the loss of grass between tree canopies. The bare intercanopy areas become highly erodible. The US Forest Service, in Bandelier National Monument for example, is studying how to restore the former ecosystem, and reduce erosion, by removing the trees.

Tree roots bind soil together, and if the soil is sufficiently shallow they act to keep the soil in place by also binding with underlying bedrock. Tree removal on steep slopes with shallow soil thus increases the risk of landslides, which can threaten people living nearby. However most deforestation only affects the trunks of trees, allowing for the roots to stay rooted, negating the landslide.

ECOLOGICAL

Deforestation results in declines in biodiversity. The removal or destruction of areas of forest cover has resulted in a degraded environment with reduced biodiversity. Forests support biodiversity, providing habitat for wildlife; moreover, forests foster medicinal conservation. With forest biotopes being irreplaceable source of new drugs (such as taxol), deforestation can destroy genetic variations (such as crop resistance) irretrievably.

Since the tropical rainforests are the most diverse ecosystems on Earth and about 80% of the world's known biodiversity could be found in tropical rainforests, removal or destruction of significant areas of forest cover has resulted in a degraded environment with reduced biodiversity.

Scientific understanding of the process of extinction is insufficient to accurately make predictions about the impact of deforestation on biodiversity. Most predictions of forestry related biodiversity loss are based on species-area models, with an underlying assumption that as forest are declines species diversity will decline similarly.

However, many such models have been proven to be wrong and loss of habitat does not necessarily lead to large scale loss of species. Species-area models are known to overpredict the number of species known to be threatened in areas where actual deforestation is ongoing, and greatly overpredict the number of threatened species that are widespread.

It has been estimated that we are losing 137 plant, animal and insect species every single day due to rainforest deforestation, which equates to 50,000 species a year. Others state that tropical rainforest deforestation is contributing to the ongoing Holocene mass extinction.

The known extinction rates from deforestation rates are very low, approximately 1 species per year from mammals and birds which extrapolates to approximately 23,000 species per year for all species.

Predictions have been made that more than 40% of the animal and plant species in Southeast Asia could be wiped out in the 21st century. Such predictions were called into question by 1995 data that show that within regions of Southeast Asia much of the original forest has been converted to monospecific plantations, but that potentially endangered species are few and tree flora remains widespread and stable.

ECONOMIC IMPACT

Damage to forests and other aspects of nature could halve living standards for the world's poor and reduce global GDP by about 7% by 2050, a major report concluded at the Convention on Biological Diversity (CBD) meeting in Bonn.

Historically utilization of forest products, including timber and fuel wood, have played a key role in human societies, comparable to the roles of water and cultivable land. Today, developed countries continue to utilize timber for building houses, and wood pulp for paper. In developing countries almost three billion people rely on wood for heating and cooking.

The forest products industry is a large part of the economy in both developed and developing countries.

Short-term economic gains made by conversion of forest to agriculture, or over-exploitation of wood products, typically leads to loss of long-term income and long term biological productivity (hence reduction in nature's services). West Africa, Madagascar, Southeast Asia and many other regions have experienced lower revenue because of declining timber harvests. Illegal logging causes billions of dollars of losses to national economies annually.

The new procedures to get amounts of wood are causing more harm to the economy and overpowers the amount of money spent by people employed in logging. According to a study, "in most areas studied, the various ventures that prompted deforestation rarely generated more than US$5 for every ton of carbon they released and frequently returned far less than US $1. " The price on the European market for an offset tied to a one-ton reduction in carbon is 23 euro (about $35).

Deforestation has been practiced by humans for tens of thousands of years before the beginnings of civilization. Fire was the first tool that allowed humans to modify the landscape. The first evidence of deforestation appears in the Mesolithic period. It was probably used to convert closed forests into more open ecosystems favourable to game animals. With the advent of agriculture, fire became the prime tool to clear land for crops.

In Europe there is little solid evidence before 7000 BC. Mesolithic foragers used fire to create openings for red deer and wild boar. In Great Britain shade tolerant species such as oak and ash are replaced in the pollen record by hazels, brambles, grasses and nettles. Removal of the forests led to decreased transpiration resulting in the formation of upland peat bogs.

Widespread decrease in elm pollen across Europe between 8400-8300 BC and 7200-7000 BC, starting in southern Europe and gradually moving north to Great Britain, may represent land clearing by fire at the onset of Neolithic agriculture.

The Neolithic period saw extensive deforestation for farming land. Stone axes were being made from about 3000 BC not just from flint, but from a wide variety of hard rocks from across Britain and North America as well. They include the noted Langdale axe industry in the English Lake District, quarries developed at Penmaenmawr in North Wales and numerous other locations.

Rough-outs were made locally near the quarries, and some were polished locally to give a fine finish. This step not only increased the mechanical strength of the axe, but also made penetration of wood easier. Flint was still used from sources such as Grimes Graves but from many other mines across Europe.

Evidence of deforestation has been found in Minoan Crete; for example the environs of the Palace of Knossos were severely deforested in the Bronze Age.

Throughout most of history, humans were hunter gatherers who hunted within forests. In most areas, such as the Amazon, the tropics, Central America, and the Caribbean, only after shortages of wood and other forest products occur are policies implemented to ensure forest resources are used in a sustainable manner.

In ancient Greece, Tjeered van Andel and co-writers summarized three regional studies of historic erosion and alluviation and found that, wherever adequate evidence exists, a major phase of erosion follows, by about 500-1,000 years the introduction of farming in the various regions of Greece, ranging from the later Neolithic to the Early Bronze Age.

The thousand years following the mid-first millennium BCE saw serious, intermittent pulses of soil erosion in numerous places. The historic silting of ports along the southern coasts of Asia Minor (*e. g.* Clarus, and the examples of Ephesus, Priene and Miletus, where harbors had to be abandoned because of the silt deposited by the Meander) and in coastal Syria during the last centuries BC.

Easter Island has suffered from heavy soil erosion in recent centuries, aggravated by agriculture and deforestation. Jared Diamond gives an extensive look into the collapse of the ancient Easter Islanders in his book *Collapse*. The disappearance of the island's trees seems to coincide with a decline of its civilization around the 17th and 18th century.

The famous silting up of the harbor for Bruges, which moved port commerce to Antwerp, also follow a period of increased settlement growth (and apparently of deforestation) in the upper river basins. In early medieval Riez in upper Provence, alluvial silt from two small rivers raised the riverbeds and widened the floodplain, which slowly buried the Roman settlement in alluvium and gradually moved new construction to higher ground; concurrently the headwater valleys above Riez were being opened to pasturage.

A typical progress trap is that cities were often built in a forested area providing wood for some industry (e. g. construction, shipbuilding, pottery). When deforestation occurs without proper replanting, local wood supplies become difficult to obtain near enough to remain competitive, leading to the city's abandonment, as happened repeatedly in Ancient Asia Minor.

The combination of mining and metallurgy often went along this self-destructive path.

Meanwhile most of the population remaining active in (or indirectly dependent on) the agricultural sector, the main pressure in most areas remained land clearing for crop and cattle farming; fortunately enough wild green was usually left standing (and partially used, e. g. to collect firewood, timber and fruits, or to graze pigs) for wildlife to remain viable, and the hunting privileges of the elite (nobility and higher clergy) often protected significant woodlands.

Major parts in the spread (and thus more durable growth) of the population were played by monastical 'pioneering' (especially by the Benedictine and Commercial orders) and some feudal lords actively attracting farmers to settle (and become tax payers) by offering relatively good legal and fiscal conditions – even when they did so to launch or encourage cities, there always was an agricultural belt around and even quite some within the walls.

When on the other hand demography took a real blow by such causes as the Black Death or devastating warfare (e. g. Genghis Khan's Mongol hordes in eastern and central Europe, Thirty Years' War in Germany) this could lead to settlements being abandoned, leaving land to be reclaimed by nature, even though the secondary forests usually lacked the original biodiversity.

From 1100 to 1500 AD significant deforestation took place in Western Europe as a result of the expanding human population. The large-scale building of wooden sailing ships by European (coastal) naval owners since the 15th century for exploration, colonisation, slave trade – and other trade on the high seas and (often related) naval warfare (the failed invasion of England by the Spanish Armada in 1559 and the battle of Lepanto 1571 are early cases of huge waste of prime timber; each of Nelson's Royal navy war ships at Trafalgar had required 6,000 mature oaks) and piracy meant that whole woody regions were over-harvested, as in Spain, where this contributed to the paradoxical weakening of the domestic economy since Columbus' discovery of America made the colonial activities (plundering, mining, cattle, plantations, trade, etc.) predominant.

In *Changes in the Land* (1983), William Cronon collected 17th century New England Englishmen's reports of increased seasonal flooding during the time that the forests were initially cleared, and it was widely believed that it was linked with widespread forest clearing upstream.

The massive use of charcoal on an industrial scale in Early Modern Europe was a new acceleration of the onslaught on western forests; even in Stuart England, the relatively primitive production of charcoal has already reached an impressive level.

For ship timbers, Stuart England was so widely deforested that it depended on the Baltic trade and looked to the untapped forests of New England to supply the need. In France, Colbert planted oak forests to supply the French navy in the future; as it turned out, as the oak plantations matured in the mid-nineteenth century, the masts were no longer required.

Norman F. Cantor's summary of the effects of late medieval deforestation applies equally well to Early Modern Europe:

Europeans had lived in the midst of vast forests throughout the earlier medieval centuries. After 1250 they became so skilled at deforestation that by 1500 AD they were running short of wood for heating and cooking.

They were faced with a nutritional decline because of the elimination of the generous supply of wild game that had inhabited the now-disappearing forests, which throughout medieval times had provided the staple of their carnivorous high-protein diet. By 1500 Europe was on the edge of a fuel and nutritional disaster, [from] which it was saved in the sixteenth century only by the burning of soft coal and the cultivation of potatoes and maize.

Specific parallels are seen in twentieth century deforestation occurring in many developing nations.

RATES OF DEFORESTATION

Global deforestation sharply accelerated around 1852. It has been estimated that about half of the earth's mature tropical forests – between 7. 5 million and 8 million km^2 (2. 9 million to 3 million sq mi) of the original 15 million to 16 million km^2 (5. 8 million to 6. 2 million sq mi) that until 1947 covered the planet – have now been cleared.

Some scientists have predicted that unless significant measures (such as seeking out and protecting old growth forests that haven't been disturbed) are taken on a worldwide basis, by 2030 there will only be ten per cent remaining, with another ten per cent in a degraded condition. 80 per cent will have been lost, and with them hundreds of thousands of irreplaceable species.

The difficulties of estimating deforestation rates are nowhere more apparent than in the widely varying estimates of rates of rainforest deforestation. Some environmental groups argue that one fifth of the world's tropical rainforest was destroyed between 1960 and 1990, that rainforests 50 years ago covered 14% of the world's land surface and have been reduced to 6%, and that all tropical forests will be gone by the year 2090.

Meanwhile, Alan Grainger of Leeds University argues that there is no credible evidence of any long-term decline in rainforest area. Bjørn Lomborg, author of the controversial book *The Skeptical Environmentalist*, claims that global forest cover has remained approximately stable since the middle of the twentieth century. Along similar lines, some have claimed that for every acre of rain forest cut down each year, more than 50 acres of new forest are growing in the tropics.

These divergent viewpoints are the result of the uncertainties in the extent of tropical deforestation. For tropical countries, deforestation estimates are very uncertain and could be in error by as much as +/- 50%, while a 2002 analysis of satellite imagery suggested that the rate of deforestation in the humid tropics (approximately 5. 8 million hectares per year) was roughly 23% lower than the most commonly quoted rates. Conversely, a new analysis of satellite images reveals that deforestation of the Amazon rainforest is twice as fast as scientists previously estimated.

Some have argued that deforestation trends may follow a Kuznets curve, which if true would nonetheless fail to eliminate the risk of irreversible loss of non-economic forest values (e. g., the extinction of species).

A 2005 report by the United Nations Food and Agriculture Organization (FAO) estimates that although the earth's total forest area continues to decrease at about 13 million hectares per year, the global rate of deforestation has recently been slowing. Still others claim that rainforests are being destroyed at an ever-quickening pace.

The London-based Rainforest Foundation notes that "the UN figure is based on a definition of forest as being an area with as little as 10% actual tree cover, which would therefore include areas that are actually savannah-like ecosystems and badly damaged forests. " Other critics of the FAO data point out that they do not distinguish between forest types, and that they are based largely on reporting from forestry departments of individual countries, which do not take into account unofficial activities like illegal logging.

Despite these uncertainties, there is agreement that destruction of rainforests remains a significant environmental problem. Up to 90% of West Africa's coastal rainforests have disappeared since 1900. In South Asia, about 88% of the rainforests have been lost. Much of what remains of the world's rainforests is in the Amazon basin, where the Amazon Rainforest covers approximately 4 million square kilometres.

The regions with the highest tropical deforestation rate between 2000 and 2005 were Central America — which lost 1. 3% of its forests each year — and tropical Asia.

In Central America, two-thirds of lowland tropical forests have been turned into pasture since 1950 and 40% of all the rainforests have been lost in the last 40 years. Brazil has lost 90-95% of its Mata Atlântica forest. Madagascar has lost 90% of its eastern rainforests.

As of 2007, less than 1% of Haiti's forests remained. Mexico, India, the Philippines, Indonesia, Thailand, Myanmar, Malaysia, Bangladesh, China, Sri Lanka, Laos, Nigeria, the Democratic Republic of the Congo, Liberia, Guinea, Ghana and the Côte d'Ivoire, have lost large areas of their rainforest. Several countries, notably Brazil, have declared their deforestation a national emergency.

CONTROLLING DEFORESTATION

KYOTO PROTOCOL

A key factor in controlling deforestation could come from the Kyoto Protocol. Reducing Emissions from Deforestation and Forest Degradation (REDD) mechanisms, which provide financial incentives for avoided deforestation, could be implemented in a future Kyoto

Protocol. At the moment, REDD is not implemented in any of the flexible mechanisms defined under the Kyoto Protocol (Clean Development Mechanism, Joint Implementation and Emissions Trading).

FARMING

New methods are being developed to farm more intensively, such as high-yield hybrid crops, greenhouse, autonomous building gardens, and hydroponics. These methods are often dependent on chemical inputs to maintain necessary yields. In cyclic agriculture, cattle are grazed on farm land that is resting and rejuvenating.

Cyclic agriculture actually increases the fertility of the soil. Intensive farming can also decrease soil nutrients by consuming at an accelerated rate the trace minerals needed for crop growth.

FOREST MANAGEMENT

Efforts to stop or slow deforestation have been attempted for many centuries because it has long been known that deforestation can cause environmental damage sufficient in some cases to cause societies to collapse. In Tonga, paramount rulers developed policies designed to prevent conflicts between short-term gains from converting forest to farmland and long-term problems forest loss would cause, while during the seventeenth and eighteenth centuries in Tokugawa, Japan, the shoguns developed a highly sophisticated system of long-term planning to stop and even reverse deforestation of the preceding centuries through substituting timber by other products and more efficient use of land that had been farmed for many centuries.

In sixteenth century Germany landowners also developed silviculture to deal with the problem of deforestation. However, these policies tend to be limited to environments with *good rainfall, no dry season* and *very young soils* (through volcanism or glaciation). This is because on older and less fertile soils trees grow too slowly for silviculture to be economic, whilst in areas with a strong dry season there is always a risk of forest fires destroying a tree crop before it matures.

In the areas where "slash-and-burn" is practiced, switching to "slash-and-char" would prevent the rapid deforestation and subsequent degradation of soils. The biochar thus created, given back to the soil, is not only a durable carbon sequestration method, but it also is an extremely beneficial amendment to the soil. Mixed with biomass it brings the creation of terra preta, one of the richest soils on the planet and the only one known to regenerate itself.

Certification of Sustainable Forest Management Practices

Certification, as provided by global certification systems such as PEFC and FSC, contributes to tackling deforestation by creating market demand for timber from sustainably managed forests. According to the United Nations Food and Agriculture Organization (FAO), "A major condition for the adoption of sustainable forest management is a demand for products that are produced sustainably and consumer willingness to pay for the higher costs entailed.

Certification represents a shift from regulatory approaches to market incentives to promote sustainable forest management. By promoting the positive attributes of forest products from sustainably managed forests, certification focuses on the demand side of environmental conservation."

REFORESTATION

In many parts of the world, especially in East Asian countries, reforestation and afforestation are increasing the area of forested lands. The amount of woodland has increased in 22 of the world's 50 most forested nations. Asia as a whole gained 1 million hectares of forest between 2000 and 2005. Tropical forest in El Salvador expanded more than 20 per cent between 1992 and 2001. Based on these trends, one study projects that global forest will increase by 10 per cent—an area the size of India—by 2050.

In the People's Republic of China, where large scale destruction of forests has occurred, the government has in the past required that every able-bodied citizen between the ages of 11 and 60 plant three to five trees per year or do the equivalent amount of work in other forest services. The government claims that at least 1 billion trees have been planted in China every year since 1982.

This is no longer required today, but March 12 of every year in China is the Planting Holiday. Also, it has introduced the Green Wall of China project, which aims to halt the expansion of the Gobi desert through the planting of trees. However, due to the large per centage of trees dying off after planting (up to 75%), the project is not very successful. There has been a 47-million-hectare increase in forest area in China since the 1970s. The total number of trees amounted to be about 35 billion and 4. 55% of China's land mass increased in forest coverage. The forest coverage was 12% two decades ago and now is 16. 55%.

An ambitious proposal for China is the Aerially Delivered Reforestation and Erosion Control System and the Proposed sahara forest project coupled with the Seawater Greenhouse

In Western countries, increasing consumer demand for wood products that have been produced and harvested in a sustainable manner is causing forest landowners and forest industries to become increasingly accountable for their forest management and timber harvesting practices.

The Arbor Day Foundation's Rain Forest Rescue programme is a charity that helps to prevent deforestation. The charity uses donated money to buy up and preserve rainforest land before the lumber companies can buy it. The Arbor Day Foundation then protects the land from deforestation.

This also locks in the way of life of the primitive tribes living on the forest land. Organizations such as Community Forestry International, Cool Earth, The Nature Conservancy, World Wide Fund for Nature, Conservation International, African Conservation Foundation and Greenpeace also focus on preserving forest habitats. Greenpeace in particular has also mapped out the forests that are still intact and published this information on the internet.

HowStuffWorks in turn has made a simpler thematic map showing the amount of forests present just before the age of man (8000 years ago) and the current (reduced) levels of forest. These maps mark the amount of afforestation required to repair the damage caused by man.

FOREST PLANTATIONS

To meet the world's demand for wood, it has been suggested by forestry writers Botkins and Sedjo that high-yielding forest plantations are suitable. It has been calculated that plantations yielding 10 cubic meters per hectare annually could supply all the timber required for international trade on 5 per cent of the world's existing forestland.

By contrast, natural forests produce about 1-2 cubic meters per hectare; therefore, 5 to 10 times more forestland would be required to meet demand. Forester Chad Oliver has suggested a forest mosaic with high-yield forest lands interpersed with conservation land.

One analysis of FAO data suggests that afforestation and reforestation projects "could reverse the global decline in woodlands within 30 years."

Reforestation through tree planting could take advantage of changing precipitation patterns due to climate change. This would be done by studying where precipitation is projected to increase and setting up reforestation projects in these locations. Areas such as Niger, Sierra Leone and Liberia are especially important candidates because they also suffer from an expanding desert (the Sahara) and decreasing biodiversity (while being important biodiversity hotspots).

MILITARY CONTEXT

While the preponderance of deforestation is due to demands for agricultural and urban use for the human population, there are some examples of military causes. One example of deliberate deforestation is that which took place in the U. S. zone of occupation in Germany after World War II. Before the onset of the Cold War defeated Germany was still considered a potential future threat rather than potential future ally.

To address this threat, attempts were made to lower German industrial potential, of which forests were deemed an element. Sources in the U. S. government admitted that the purpose of this was the "ultimate destruction of the war potential of German forests. " As a consequence of the practice of clear-felling, deforestation resulted which could "be replaced only by long forestry development over perhaps a century. "

War can also be a cause of deforestation, either deliberately such as through the use of Agent Orange during the Vietnam War where, together with bombs and bulldozers, it contributed to the destruction of 44 per cent of the forest cover, or inadvertently such as in the 1945 Battle of Okinawa where bombardment and other combat operations reduced the lush tropical landscape into "a vast field of mud, lead, decay and maggots".

Chapter 12

Global Warming

Global warming is the increase in the average temperature of the Earth's near-surface air and oceans since the mid-20th century and its projected continuation. Global surface temperature increased 0. 74 ± 0. 18 °C (1. 33 ± 0. 32 °F) during the last century. The Intergovernmental Panel on Climate Change (IPCC) concludes that increasing greenhouse gas concentrations resulting from human activity such as fossil fuel burning and deforestation are responsible for most of the observed temperature increase since the middle of the 20th century.

The IPCC also concludes that natural phenomena such as solar variation and volcanoes produced most of the warming from pre-industrial times to 1950 and had a small cooling effect afterward. These basic conclusions have been endorsed by more than 45 scientific societies and academies of science, including all of the national academies of science of the major industrialized countries.

Climate model projections summarized in the latest IPCC report indicate that the global surface temperature will probably rise a further 1. 1 to 6. 4 °C (2. 0 to 11. 5 °F) during the twenty-first century. The uncertainty in this estimate arises from the use of models with differing sensitivity to greenhouse gas concentrations and the use of differing estimates of future greenhouse gas emissions.

Some other uncertainties include how warming and related changes will vary from region to region around the globe. Most studies focus on the period up to the year 2100. However, warming is expected to continue beyond 2100 even if emissions stop, because of the large heat capacity of the oceans and the long lifetime of carbon dioxide in the atmosphere.

An increase in global temperature will cause sea levels to rise and will change the amount and pattern of precipitation, probably including expansion of subtropical deserts. The continuing retreat of glaciers, permafrost and sea ice is expected, with the Arctic region being particularly affected. Other likely effects include shrinkage of the Amazon rainforest and Boreal forests, increases in the intensity of extreme weather events, species extinctions, and changes in agricultural yields.

Political and public debate continues regarding climate change, and what actions (if any) to take in response. The available options are mitigation to reduce further emissions; adaptation to reduce the damage caused by warming; and, more speculatively, geoengineering to reverse global warming. Most national governments have signed and ratified the Kyoto Protocol aimed at reducing greenhouse gas emissions.

TEMPERATURE CHANGES

The most commonly cited indication of global warming is the trend in globally averaged temperature near the Earth's surface. Expressed as a linear trend, this temperature rose by 0. 74°C ±0. 18°C over the period 1906-2005. The rate of warming over the last 50 years of that period was almost double that for the period as a whole (0. 13°C ±0. 03°C per decade, versus 0. 07°C ± 0. 02°C per decade).

The urban heat island effect is estimated to account for about 0. 002 °C of warming per decade since 1900. Temperatures in the lower troposphere have increased between 0. 12 and 0. 22 °C (0. 22 and 0. 4 °F) per decade since 1979, according to satellite temperature measurements. Temperature is believed to have been relatively stable over the one or two thousand years before 1850, with regionally-varying fluctuations such as the Medieval Warm Period or the Little Ice Age.

Based on estimates by NASA's Goddard Institute for Space Studies, 2005 was the warmest year since reliable, widespread instrumental measurements became available in the late 1800s, exceeding the previous record set in 1998 by a few hundredths of a degree.

Estimates prepared by the World Meteorological Organization and the Climatic Research Unit concluded that 2005 was the second warmest year, behind 1998. Temperatures in 1998 were unusually warm because the strongest El Niño in the past century occurred during that year.

Temperature changes vary over the globe. Since 1979, land temperatures have increased about twice as fast as ocean temperatures (0. 25 °C per decade against 0. 13 °C per decade). Ocean temperatures increase more slowly than land temperatures because of the larger effective heat capacity of the oceans and because the ocean loses more heat by evaporation.

The Northern Hemisphere warms faster than the Southern Hemisphere because it has more land and because it has extensive areas of seasonal snow and sea-ice cover subject to the ice-albedo feedback. Although more greenhouse gases are emitted in the Northern than Southern Hemisphere this does not contribute to the difference in warming because the major greenhouse gases persist long enough to mix between hemispheres.

The thermal inertia of the oceans and slow responses of other indirect effects mean that climate can take centuries or longer to adjust to changes in forcing. Climate commitment studies indicate that even if greenhouse gases were stabilized at 2000 levels, a further warming of about 0. 5 °C (0. 9 °F) would still occur.

Radiative Forcing

External forcing is a term used in climate science for processes external to the climate system (though not necessarily external to Earth). Climate responds to several types of external forcing, such as changes in greenhouse gas concentrations, changes in solar luminosity, volcanic eruptions, and variations in Earth's orbit around the Sun.

Attribution of recent climate change focuses on the first three types of forcing. Orbital cycles vary slowly over tens of thousands of years and thus are too gradual to have caused the temperature changes observed in the past century.

Greenhouse Gases

The greenhouse effect is the process by which absorption and emission of infrared radiation by gases in the atmosphere warm a planet's lower atmosphere and surface. It was discovered by Joseph Fourier in 1824 and was first investigated quantitatively by Svante Arrhenius in 1896.

Existence of the greenhouse effect as such is not disputed, even by those who do not agree that the recent temperature increase is attributable to human activity. The question is instead how the strength of the greenhouse effect changes when human activity increases the concentrations of greenhouse gases in the atmosphere.

Naturally occurring greenhouse gases have a mean warming effect of about 33 °C (59 °F). The major greenhouse gases are water vapour (not including clouds), which causes about 36–70 per cent of the greenhouse effect; carbon dioxide (CO_2), which causes 9–26 per cent; methane (CH_4), which causes 4–9 per cent; and ozone (O_3), which causes 3–7 per cent.

Human activity since the Industrial Revolution has increased the amount of greenhouse gases in the atmosphere, leading to increased radiative forcing from CO_2, methane, tropospheric ozone, CFCs and nitrous oxide. The concentrations of CO_2 and methane have increased by 36% and 148% respectively since the mid-1700s. These levels are considerably higher than at any time during the last 650,000 years, the period for which reliable data has been extracted from ice cores. Less direct geological evidence indicates that CO_2 values this high were last seen approximately 20 million years ago.

Fossil fuel burning has produced about three-quarters of the increase in CO_2 from human activity over the past 20 years. Most of the rest is due to land-use change, particularly deforestation.

CO_2 concentrations are continuing to rise due to burning of fossil fuels and land-use change. The future rate of rise will depend on uncertain economic, sociological, technological, and natural developments. Accordingly, the IPCC Special Report on Emissions Scenarios gives a wide range of future CO_2 scenarios, ranging from 541 to 970 ppm by the year 2100. Fossil fuel reserves are sufficient to reach these levels and continue emissions past 2100 if coal, tar sands or methane clathrates are extensively exploited.

The destruction of stratospheric ozone by chlorofluorocarbons is sometimes mentioned in relation to global warming. Although there are a few areas of linkage, the relationship between the two is not strong. Reduction of stratospheric ozone has a cooling influence, but substantial ozone depletion did not occur until the late 1970s. Tropospheric ozone contributes to surface warming.

AEROSOLS AND SOOT

Global dimming, a gradual reduction in the amount of global direct irradiance at the Earth's surface, has partially counteracted global warming from 1960 to the present. The main cause of this dimming is aerosols produced by volcanic activity and emissions of pollutants such as sulfur dioxide.

These aerosols exert a cooling effect by increasing the reflection of incoming sunlight. James Hansen and colleagues have proposed that the effects of the products of fossil fuel combustion—CO_2 and aerosols—have largely offset one another in recent decades, so that net warming has been driven mainly by non-CO_2 greenhouse gases.

In addition to their direct effect by scattering and absorbing solar radiation, aerosols have indirect effects on the radiation budget. Sulfate aerosols act as cloud condensation nuclei and thus lead to clouds that have more and smaller cloud droplets. These clouds reflect solar radiation more efficiently than clouds with fewer and larger droplets.

This effect also causes droplets to be of more uniform size, which reduces growth of raindrops by collision-coalescence. Clouds modified by pollution have been shown to produce less drizzle, making the cloud brighter and more reflective to incoming sunlight, especially in the near-infrared part of the spectrum.

Soot may cool or warm, depending on whether it is airborne or deposited. Atmospheric soot aerosols directly absorb solar radiation, which heats the atmosphere and cools the surface. Regionally (but not globally), as much as 50% of surface warming due to greenhouse gases may be masked by atmospheric brown clouds.

When deposited, especially on glaciers or on ice in arctic regions, the lower surface albedo can also directly heat the surface. The influences of aerosols, including black carbon, are most pronounced in the tropics and sub-tropics, particularly in Asia, while the effects of greenhouse gases are dominant in the extratropics and southern hemisphere.

SOLAR VARIATION

Variations in solar output have been the cause of past climate changes. Although solar forcing is generally thought to be too small to account for a significant part of global warming in recent decades, a few studies disagree, such as a recent phenomenological analysis that indicates the contribution of solar forcing may be underestimated.

Greenhouse gases and solar forcing affect temperatures in different ways. While both increased solar activity and increased greenhouse gases are expected to warm the troposphere, an increase in solar activity should warm the stratosphere while an increase in greenhouse gases should cool the stratosphere.

Observations show that temperatures in the stratosphere have been steady or cooling since 1979, when satellite measurements became available. Radiosonde (weather balloon) data from the pre-satellite era show cooling since 1958, though there is greater uncertainty in the early radiosonde record. A related hypothesis, proposed by Henrik Svensmark, is that magnetic activity of the sun deflects cosmic rays that may influence the generation of cloud condensation nuclei and thereby affect the climate. Other research has found no relation between warming in recent decades and cosmic rays. The influence of cosmic rays on cloud cover is about a factor of 100 lower than needed to explain the observed changes in clouds or to be a significant contributor to present-day climate change.

GLOBAL WARMING AWARENESS

The average facade temperature of the globe has augmented more than 1 degree Fahrenheit since 1900 and the speed of warming has been almost three folds the century long average since 1970. This increase in earth's average temperature is called Global warming. More or less all specialists studying the climate record of the earth have the same opinion now that human actions, mainly the discharge of green house gases from smokestacks, vehicles, and burning forests, are perhaps the leading power driving the fashion.

Fig. Melting of Glaciers are the Effect of Global Warming

The gases append to the planet's normal greenhouse effect, permitting sunlight in, but stopping some of the ensuing heat from radiating back to space. Based on the study on past climate shifts, notes of current situations, and computer simulations, many climate scientists say that lacking of big curbs in greenhouse gas discharges, the 21st century might see temperatures rise of about 3 to 8 degrees, climate patterns piercingly shift, ice sheets contract and seas rise several feet.

With the probable exemption of one more world war, a huge asteroid, or a fatal plague, global warming may be the only most danger to our planet earth.

Global Warming Causes

As said, the major cause of global warming is the emission of green house gases like carbon dioxide, methane, nitrous oxide etc into the atmosphere.

The major source of carbon dioxide is the power plants. These power plants emit large amounts of carbon dioxide produced from burning of fossil fuels for the purpose of electricity generation. About twenty per cent of carbon dioxide emitted in the atmosphere comes from burning of gasoline in the engines of the vehicles. This is true for most of the developed countries. Buildings, both commercial and residential represent a larger source of global warming pollution than cars and trucks.

Building of these structures require a lot of fuel to be burnt which emits a large amount of carbon dioxide in the atmosphere. Methane is more than 20 times as effectual as CO2 at entrapping heat in the atmosphere. Methane is obtained from resources such as rice paddies, bovine flatulence, bacteria in bogs and fossil fuel manufacture. When fields are flooded, anaerobic situation build up and the organic matter in the soil decays, releasing methane to the atmosphere.

The main sources of nitrous oxide include nylon and nitric acid production, cars with catalytic converters, the use of fertilizers in agriculture and the burning of organic matter. Another cause of global warming is deforestation that is caused by cutting and burning of forests for the purpose of residence and industrialization.

Global Warming is Inspiring Scientists to Fight for Awareness

Scientists all over the world are making predictions about the ill effects of Global warming and connecting some of the events that have taken place in the pat few decades as an alarm of global warming. The effect of global warming is increasing the average temperature of the earth.

A rise in earth's temperatures can in turn root to other alterations in the ecology, including an increasing sea level and modifying the quantity and pattern of rainfall. These modifications may boost the occurrence and concentration of severe climate events, such as floods, famines, heat waves, tornados, and twisters. Other consequences may comprise of higher or lower agricultural outputs, glacier melting, lesser summer stream flows, genus extinctions and rise in the ranges of disease vectors.

As an effect of global warming species like golden toad, harlequin frog of Costa Rica has already become extinct. There are number of species that have a threat of disappearing soon as an effect of global warming. As an effect of global warming various new diseases have emerged lately.

These diseases are occurring frequently due to the increase in earths average temperature since the bacteria can survive better in elevated temperatures and even multiplies faster when the conditions are favorable. The global warming is extending the distribution of mosquitoes due to the increase in humidity levels and their frequent growth in warmer atmosphere. Various diseases due to ebola, hanta and machupo virus are expected due to warmer climates. The marine life is also very sensitive to the increase in temperatures.

The effect of global warming will definitely be seen on some species in the water. A survey was made in which the marine life reacted significantly to the changes in water temperatures. It is expected that many species will die off or become extinct due to the increase in the temperatures of the water, whereas various other species, which prefer warmer waters, will increase tremendously.

Perhaps the most disturbing changes are expected in the coral reefs that are expected to die off as an effect of global warming. The global warming is expected to cause irreversible changes in the ecosystem and the behaviour of animals.

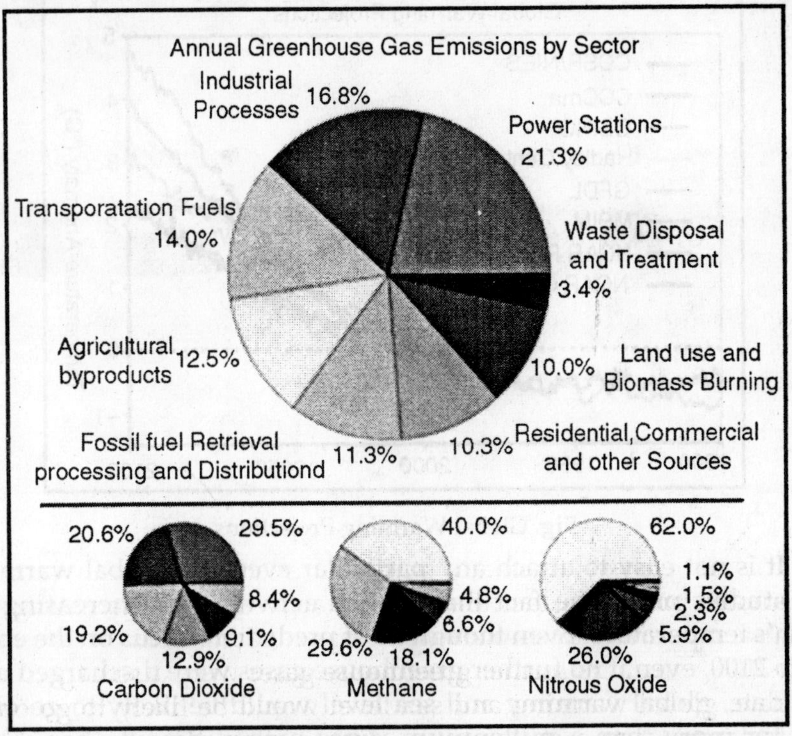

Fig. Annual Green House Gas Emission

A group of scientists have recently reported on the surprisingly speedy rise in the discharge of carbon and methane release from frozen tundra in Siberia, now starting to melt because of human cause increases in earth' 's temperature.

The scientists tell us that the tundra is in danger of melting holds an amount of extra global warming pollution that is equivalent to the net amount that is previously in the earth's atmosphere. Likewise, earlier one more team of scientists reported that the in a single year Greenland witnessed 32 glacial earthquakes between 4. 6 and 5. 1 on the Richter scale.

This is a disturbing sign and points that a huge destabilization that may now be in progress deep within the second biggest accretion of ice on the planet. This ice would be enough to raise sea level 20 feet worldwide if it broke up and slipped into the sea. Each day passing brings yet new proof that we are now in front of a global emergency, a climate emergency that needs instant action to piercingly decrease carbon dioxide emissions worldwide in order to turn down the earth's rising temperatures and avoid any catastrophe.

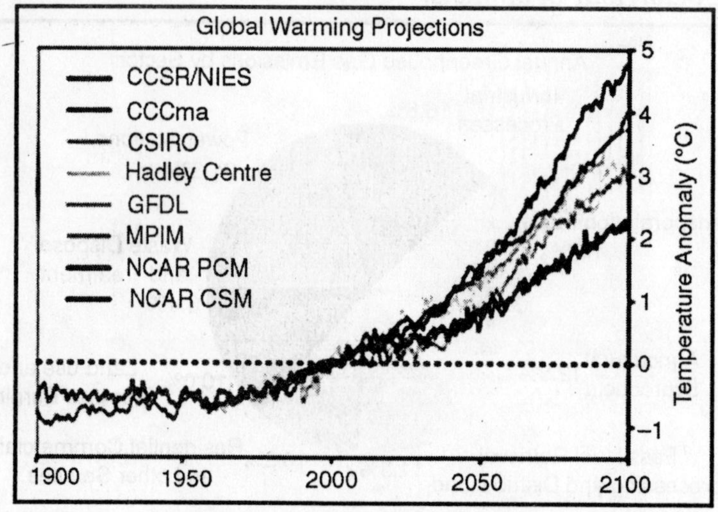

Fig. Global Warming Projections

It is not easy to attach any particular events to global warming, but studies prove the fact that human activities are increasing the earth's temperature. Even though most predictions focus on the epoch up to 2100, even if no further greenhouse gases were discharged after this date, global warming and sea level would be likely to go on to rise for more than a millennium, since carbon dioxide has a long average atmospheric life span.

FIGHT GLOBAL WARMING

Many efforts are being made by various nations to cut down the rate of global warming. One such effort is the Kyoto agreement that has been made between various nations to reduce the emissions of various green house gases. Also many non profit organizations are working for the cause. Al Gore was one of the foremost U. S. politicians to heave an alarm about the hazards of global warming.

He has produced a significantly acclaimed documentary movie called "An Inconvenient Truth," and written a book that archives his advice that Earth is dashing toward an immensely warmer future. Al Gore, the former vice president of United States has given various speeches to raise an awareness of global warming. He has warned people about the ill effects of Global warming and its remedies.

But an interesting side of the global warming episode is that there are people who do not consider global warming as something that is creating a problem. Skeptics of global warming think that global warming is not an ecological trouble.

Fig. Benifits of Global Warming

According to the global warming skeptics, the recent enhancement in the earth's average temperature is no reason for alarm. According to them earth's coastlines and polar ice caps are not at a risk of vanishing. Global warming skeptics consider that the weather models used to establish global warming and to forecast its impacts are distorted.

According to the models, if calculations are made the last few decades must have been much worse as compared to actually happened to be. Most of the global warming skeptics believe that the global warming is not actually occurring. They stress on the fact the climatic conditions vary because of volcanism, the obliquity cycle, changes in solar output, and internal variability.

Also the warming can be due to the variation in cloud cover, which in turn is responsible for the temperatures on the earth. The variations are also a result of cosmic ray flux that is modulated by the solar magnetic cycles.

Global Warming Skeptics

The global warming skeptics are of the view that the global warming is a good phenomenon and should not be stopped. There are various benefits of global warming according to them. According to the skeptics, the global warming will increase humidity in tropical deserts.

Also the higher levels of carbon dioxide in the atmosphere trigger plant growth. As predicted, due to the global warming the sea levels will rise. But this can be readily adapted. Another argument of global warming skeptics is that earth has been warmer than today as seen in its history. The thought is that global warming is nothing to get afraid of because it just takes us back to a more natural set of environment of the past. Animals and plants appeared to do just fine in those eras of warm climate on the earth. According to few skeptics, the present chilly climate on the earth is an abnormality when judged over the geographical scale. Over geologic time, the earth's mean temperature is 22 degrees C, as compared to today's 15. 5 degrees C.

Fig. Result of Global Warming

A positive feedback is a process that amplifies some change. Thus, when a warming trend results in effects that induce further warming, the result is a positive feedback; when the warming results in effects that reduce the original warming, the result is a negative feedback.

The main positive feedback in global warming involves the tendency of warming to increase the amount of water vapour in the atmosphere. The main negative feedback in global warming is the effect of temperature on emission of infrared radiation: as the temperature of a body increases, the emitted radiation increases with the fourth power of its absolute temperature.

Water Vapour Feedback

If the atmosphere is warmed, the saturation vapour pressure increases, and the amount of water vapour in the atmosphere will tend to increase. Since water vapour is a greenhouse gas, the increase in water vapour content makes the atmosphere warm further; this warming causes the atmosphere to hold still more water vapour (a positive feedback), and so on until other processes stop the feedback loop. The result is a much larger greenhouse effect than that due to CO_2 alone. Although this feedback process causes an increase in the absolute moisture content of the air, the relative humidity stays nearly constant or even decreases slightly because the air is warmer.

Cloud Feedback

Warming is expected to change the distribution and type of clouds. Seen from below, clouds emit infrared radiation back to the surface, and so exert a warming effect; seen from above, clouds reflect sunlight and emit infrared radiation to space, and so exert a cooling effect. Whether the net effect is warming or cooling depends on details such as the type and altitude of the cloud, details that are difficult to represent in climate models.

Lapse Rate

The atmosphere's temperature decreases with height in the troposphere. Since emission of infrared radiation varies with the fourth power of temperature, longwave radiation escaping to space from the relatively cold upper atmosphere is less than that emitted toward the ground from the lower atmosphere. Thus, the strength of the greenhouse effect depends on the atmosphere's rate of temperature decrease with height.

Both theory and climate models indicate that global warming will reduce the rate of temperature decrease with height, producing a negative *lapse rate feedback* that weakens the greenhouse effect. Measurements of the rate of temperature change with height are very sensitive to small errors in observations, making it difficult to establish whether the models agree with observations.

Ice-albedo Feedback

When ice melts, land or open water takes its place. Both land and open water are on average less reflective than ice and thus absorb more solar radiation. This causes more warming, which in turn causes more melting, and this cycle continues.

Arctic Methane Release

Warming is also the triggering variable for the release of methane from sources both on land and on the deep ocean floor, making both of these possible feedback effects. Thawing permafrost, such as the frozen peat bogs in Siberia, creates a positive feedback due to the release of CO_2 and methane.

Reduced Absorption of CO_2 by the Oceans

Ocean ecosystems' ability to sequester carbon is expected to decline as the oceans warm. This is because warming reduces the nutrient levels of the mesopelagic zone (about 200 to 1000 m deep), which limits the growth of diatoms in favor of smaller phytoplankton that are poorer biological pumps of carbon.

CLIMATE MODELS

The main tools for projecting future climate changes are computer models of the climate. These models are based on physical principles including fluid dynamics and radiative transfer. Although they attempt to include as many processes as possible, simplifications of the actual climate system are inevitable because of the constraints of available computer power and limitations in knowledge of the climate system.

All modern climate models include an atmospheric model that is coupled to an ocean model and models for ice cover on land and sea. Some models also include treatments of chemical and biological processes. These models project a warmer climate due to increasing levels of greenhouse gases.

Although much of the variation in model outcomes depends on the greenhouse gas emissions used as inputs, the temperature effect of a specific greenhouse gas concentration (climate sensitivity) varies depending on the model used. The representation of clouds is one of the main sources of uncertainty in present-generation models.

Global climate model projections of future climate most often have used estimates of greenhouse gas emissions from the IPCC Special Report on Emissions Scenarios (SRES). In addition to human-caused emissions, some models also include a simulation of the carbon cycle; this generally shows a positive feedback, though this response is uncertain. Some observational studies also show a positive feedback.

Including uncertainties in future greenhouse gas concentrations and climate sensitivity, the IPCC anticipates a warming of 1. 1 °C to 6. 4 °C (2. 0 °F to 11. 5 °F) by the end of the 21st century, relative to 1980–1999. A 2008 paper predicts that the global temperature may not increase during the next decade because short-term natural fluctuations may temporarily outweigh greenhouse gas-induced warming.

Models are also used to help investigate the causes of recent climate change by comparing the observed changes to those that the models project from various natural and human-derived causes. Although these models do not unambiguously attribute the warming that occurred from approximately 1910 to 1945 to either natural variation or human effects, they do indicate that the warming since 1975 is dominated by man-made greenhouse gas emissions.

Current climate models produce a good match to observations of global temperature changes over the last century, but do not simulate all aspects of climate. The physical realism of models is tested by examining their ability to simulate current or past climates.

While a 2007 study by David Douglass and colleagues found that the models did not accurately predict observed changes in the tropical troposphere, a 2008 paper published by a 17-member team led by Ben Santer noted errors in the Douglass study, and found instead that the models and observations were not statistically different.

Not all effects of global warming are accurately predicted by the climate models used by the IPCC. For example, observed Arctic shrinkage has been faster than that predicted.

ATTRIBUTED AND EXPECTED EFFECTS

Environmental

It usually is impossible to connect specific weather events to global warming. Instead, global warming is expected to cause changes in the overall distribution and intensity of events, such as changes to the frequency and intensity of heavy precipitation.

Broader effects are expected to include glacial retreat, Arctic shrinkage, and worldwide sea level rise. Other effects may include changes in crop yields, addition of new trade routes, species extinctions, and changes in the range of disease vectors.

Some effects on both the natural environment and human life are, at least in part, already being attributed to global warming. A 2001 report by the IPCC suggests that glacier retreat, ice shelf disruption such as that of the Larsen Ice Shelf, sea level rise, changes in rainfall patterns, and increased intensity and frequency of extreme weather events are attributable in part to global warming. Other expected effects include water scarcity in some regions and increased precipitation in others, changes in mountain snowpack, and adverse health effects from warmer temperatures.

Social and economic effects of global warming may be exacerbated by growing population densities in affected areas. Temperate regions are projected to experience some benefits, such as fewer cold-related deaths. A summary of probable effects and recent understanding can be found in the report made for the IPCC Third Assessment Report by Working Group II.

The newer IPCC Fourth Assessment Report summary reports that there is observational evidence for an increase in intense tropical cyclone activity in the North Atlantic Ocean since about 1970, in correlation with the increase in sea surface temperature, but that the detection of long-term trends is complicated by the quality of records prior to routine satellite observations. The summary also states that there is no clear trend in the annual worldwide number of tropical cyclones.

Additional anticipated effects include sea level rise of 0. 18 to 0. 59 meters (0. 59 to 1. 9 ft) in 2090-2100 relative to 1980-1999, repercussions to agriculture, possible slowing of the thermohaline circulation, reductions in the ozone layer, increasingly intense (but less frequent) hurricanes and extreme weather events, lowering of ocean pH, oxygen depletion in the oceans, and the spread of diseases such as malaria and dengue fever, as well as Lyme disease, hantavirus infections, bubonic plague, and cholera.

One study predicts 18% to 35% of a sample of 1,103 animal and plant species would be extinct by 2050, based on future climate projections. However, few mechanistic studies have documented extinctions due to recent climate change and one study suggests that projected rates of extinction are uncertain.

Increased atmospheric CO_2 increases the amount of CO_2 dissolved in the oceans. CO_2 dissolved in the ocean reacts with water to form carbonic acid, resulting in ocean acidification. Ocean surface pH is estimated to have decreased from 8. 25 near the beginning of the industrial era to 8. 14 by 2004, and is projected to decrease by a further 0. 14 to 0. 5 units by 2100 as the ocean absorbs more CO_2. Since organisms and ecosystems are adapted to a narrow range of pH, this raises extinction concerns, directly driven by increased atmospheric CO_2, that could disrupt food webs and impact human societies that depend on marine ecosystem services.

Heat and carbon dioxide trapped in the oceans may still take hundreds years to be re-emitted, even after greenhouse gas emissions are eventually reduced.

ECONOMIC

The IPCC reports the aggregate net economic costs of damages from climate change globally (discounted to the specified year). In 2005, the average social cost of carbon from 100 peer-reviewed estimates is US$12 per tonne of CO_2, but range -$3 to $95/t$CO_2$. The IPCC's gives these cost estimates with the caveats, "Aggregate estimates of costs mask significant differences in impacts across sectors, regions and populations and *very likely* underestimate damage costs because they cannot include many non-quantifiable impacts. "

One widely publicized report on potential economic impact is the Stern Review, written by Sir Nicholas Stern. It suggests that extreme weather might reduce global gross domestic product by up to one per cent, and that in a worst-case scenario global per capita consumption could fall by the equivalent of 20 per cent.

The response to the Stern Review was mixed. The Review's methodology, advocacy and conclusions were criticized by several economists, including Richard Tol, Gary Yohe, Robert Mendelsohn and William Nordhaus. Economists that have generally supported the Review include Terry Barker, William Cline, and Frank Ackerman. According to Barker, the costs of mitigating climate change are 'insignificant' relative to the risks of unmitigated climate change.

According to United Nations Environment Programme (UNEP), economic sectors likely to face difficulties related to climate change include banks, agriculture, transport and others. Developing countries dependent upon agriculture will be particularly harmed by global warming.

RESPONSES TO GLOBAL WARMING

The broad agreement among climate scientists that global temperatures will continue to increase has led some nations, states, corporations and individuals to implement responses. These responses to global warming can be divided into mitigation of the causes and effects of global warming, adaptation to the changing global environment, and geoengineering to reverse global warming.

Mitigation

Mitigation of global warming is accomplished through reductions in the rate of anthropogenic greenhouse gas release. Models suggest that mitigation can quickly begin to slow global warming, but that temperatures will appreciably decrease only after several centuries. The world's primary international agreement on reducing greenhouse gas emissions is the Kyoto Protocol, an amendment to the UNFCCC negotiated in 1997. The Protocol now covers more than 160 countries and over 55 per cent of global greenhouse gas emissions. As of June 2009, only the United States, historically the world's largest emitter of greenhouse gases, has refused to ratify the treaty. The treaty expires in 2012. International talks began in May 2007 on a future treaty to succeed the current one. UN negotiations are now gathering pace in advance of a meeting in Copenhagen in December 2009.

Many environmental groups encourage individual action against global warming, as well as community and regional actions. Others have suggested a quota on worldwide fossil fuel production, citing a direct link between fossil fuel production and CO_2 emissions. There has also been business action on climate change, including efforts to improve energy efficiency and limited moves towards use of alternative fuels. In January 2005 the European Union introduced its European Union Emission Trading Scheme, through which companies in conjunction with government agree to cap their emissions or to purchase credits from those below their allowances. Australia announced its Carbon Pollution Reduction Scheme in 2008. United States President Barack Obama has announced plans to introduce an economy-wide cap and trade scheme.

The IPCC's Working Group III is responsible for crafting reports on mitigation of global warming and the costs and benefits of different approaches. The 2007 IPCC Fourth Assessment Report concludes that no one technology or sector can be completely responsible for mitigating future warming. They find there are key practices and technologies in various sectors, such as energy supply, transportation, industry, and agriculture, that should be implemented to reduced global emissions. They estimate that stabilization of carbon dioxide equivalent between 445 and 710 ppm by 2030 will result in between a 0. 6 per cent increase and three per cent decrease in global gross domestic product.

Adaptation

A wide variety of measures have been suggested for adaptation to global warming. These range from the trivial, such as the installation of air-conditioning equipment, up to major infrastructure projects, such as abandonment of settlements threatened by sea level rise. Measures including water conservation, water rationing, adaptive agricultural practices, construction of flood defences, changes to medical care, and interventions to protect threatened species have all been suggested. A wide-ranging study of the possible opportunities for adaptation of infrastructure has been published by the Institute of Mechanical Engineers.

Geoengineering

Geoengineering is the deliberate modification of Earth's natural environment on a large scale to suit human needs. An example is greenhouse gas remediation, which removes greenhouse gases from the atmosphere, usually through carbon sequestration techniques such as carbon dioxide air capture. Solar radiation management reduces insolation, such as by the addition of stratospheric sulfur aerosols. No large-scale geoengineering projects have yet been undertaken.

ECONOMIC AND POLITICAL DEBATE

Increased publicity of the scientific findings surrounding global warming has resulted in political and economic debate. Poor regions, particularly Africa, appear at greatest risk from the projected effects of global warming, while their emissions have been small compared to the developed world. The exemption of developing country exempt from Kyoto Protocol restrictions have been used as part of a rationale for non-ratification by the U. S. and criticism from Austria.

In 2007-2008 the Gallup polls surveyed 127 countries. A third of the world population is unaware of the causes and issues of global warming, developing countries more unaware than developed, Africa the lowest. However, awareness does not equate to belief that global warming is a result of human activities.

Of those who are aware, Latin America leads in belief that temperature changes are a result of human activities; Africa, parts of Asia and the Middle East, a few from the Former Soviet Union lead in the opposite. In the western world, the concept and the appropriate responses are contested. Nick Pidgeon, professor at Cardiff University, describe "results show the different stages of engagement about global warming on each side of the Atlantic" where Europe debates the appropriate responses while the United States debates whether climate change is happening.

International debates weigh the benefits of limiting industrial emissions of greenhouse gases against the costs that such changes would entail. Using economic incentives, alternative and renewable energy have been promoted to reduce emissions while building infrastructure. Business-centered organizations, conservative commentators, and companies such as the Competitive Enterprise Institute and ExxonMobil have downplayed IPCC climate change scenarios, funded scientists who disagree with the scientific consensus, and provided their own projections of the economic cost of stricter controls. Environmental organizations and public figures have emphasized changes in the current climate and the risks they entail, while promoting adaptation to changes in infrastructural needs and emissions reductions. Some fossil fuel companies have scaled back their efforts in recent years, or called for policies to reduce global warming.

Another point of contention is the degree to which emerging economies such as India and China should be expected to constrain their emissions. The U. S. contends that if it must bear the cost of reducing emissions, then China should do the same since China's gross national CO_2 emissions now exceeds those of the U. S. However, China has contended that it is less of an obligated to reduce emissions since its per capita responsibility and per capita emissions are less that of the U. S. India, also exempt, has made similar contentions.

Chapter 13

Effects of Global Warming

The occurrence of unusual cold wave conditions this year throughout the globe when environmentalists voicing their concerns about the human-led global warming had predicted that the rise in carbon dioxide level in the atmosphere would result in shorter winters with no significant dip in the mercury?

Was this winter an exception to the rule or is it simply following a trend? After all, studies conducted by a small group of 'sceptic' scientists reveal that global warming has been waning since 2001. Latest studies supported by satellite data cast doubt on the climate fears propounded by environmentalists supported by the UN Intergovernmental Panel on Climate Change (IPCC).

Satellite measurements available since 1979 show no warming in the southern hemisphere and the trend in the northern hemisphere appears to have waned since 2001. In August 2007, the UK Met Office acknowledged that obvious global warming had stopped. Paleo-climate scientist Bob Carter testifying before the US Senate Committee on Environment and Public Works has noted that the accepted global average temperature statistics used by IPCC show no ground-based warming has occurred since 1998.

A research led by David Bromwich, Professor of Atmospheric Science in the Department of Geography at Ohio State University and researchers with the Byrd Polar Research Centre at Ohio State University shows that during the late 20th century, the temperature in Antarctica did not rise to the level predicted by many global warming models. According to UN scientist Madhav L. Khandekar, a retired Environment Canada scientist and an expert IPCC reviewer in 2007, the recent worldwide analysis of ocean surface temperatures shows that sea surface temperatures over world oceans are slowly declining since mid-1998.

While the level of carbon dioxide in the atmosphere is steadily rising from 280 ppm and might reach 560 ppm by 2100 as predicted by IPCC, the world's average temperature, instead of following a steep upward gradient, is actually plunging after a period of upward trend. However, the IPCC is not coming out publicly with the truth surrounding the correlation between rise in carbon dioxide level in the atmosphere and its possible consequence on global warming, if any. A study by researchers of the Atmospheric Science Group, Department of Mathematical Science, at the University of Wisconsin, found that global warming in the last century was linked to natural causes.

The Royal Meteorological Institute at Brussels in its report last year said that not carbon dioxide but the most important greenhouse gas was water vapour; it was responsible for 75 per cent of the greenhouse effect. According to Belgian climate scientist Lu Debontridder, the warm winters of the last few years in Belgium are simply due to the North-Atlantic oscillation that has absolutely nothing to do with carbon dioxide. A study published in Science last September found that contrary to past inferences from ice core records, carbon dioxide did not cause the end of the last ice age.

According to the same study, deep-sea temperatures warmed about 1,300 years before the tropical surface ocean and well before the rise in atmospheric carbon dioxide. USC geologist Lowell Scot, the lead author of the study, said that the climate dynamics are much more complex than simply saying that carbon dioxide rises and the temperature warms.

The IPCC climate model is based on the assumption that increased warming would cause more rainfall that would produce more clouds on the higher reaches of the atmosphere. Since high clouds have a net warming effect this would cause more warming, more rainfall and the cycle will continue. It is this positive feedback that causes the UN climate models to predict a temperature rise in the range of 2. 5 degree Celsius to 4. 7 degree Celsius due to rise in the level of carbon dioxide to 560 ppm. Dr Roy Spencer along with researchers at the University of Alabama Huntsville and Lawrence Livermore National Laboratory, California, after observing the temperatures, clouds and rainfall reported that warming is actually associated with fewer high clouds.

There is no data to support the theory that more rainfall will produce more high-altitude clouds.

The mainstream media seems to be purposely ignoring the bulk of the findings by renowned researchers throughout the globe that the current global warming fear attributed solely to carbon dioxide rise is utterly unfounded. Why is the IPCC, which has been blamed for relying on climate models based on wrong assumptions, continuing with its false prophecy?

Is there more to it than what meets the eye? Has the politics of carbon trade got anything to do with it? Critics say that carbon trading as propounded by IPCC, as a mean to combat global warming is a smokescreen.

It will allow corporate polluters in rich countries to evade their emission reduction obligations at home by buying up and trading carbon emission quotas and credits from other countries, projects or industries. It is meant to create further global economic disparities by robbing the poor of their rights while the rich will manage to extract maximum benefit from the mechanism.

While some would call global warming a theory, others would call it a proven set of facts. Opinions differ vehemently. Let us consider global warming to be a both a premise that the environment of the world as we know it is slowly, but very surely increasing in overall air and water temperature, and a promise that if whatever is causing this trend is not interrupted or challenged life on earth will dynamically be affected.

The prevailing counter opinion is that all that is presently perceived to be global warming is simply the result of a normal climactic swing in the direction of increased temperature. Most proponents of this global warming ideology have definitive social and financial interests in these claims.

Global warming is real. It is not the result of a natural climatic adjustment. It is a quantifiable set of environmental results that are in addition to any normal changes in climate. That is why the effects of global warming have catastrophic potential. Global warming is the straw that breaks the camel's back. It is the difference between a category three hurricane and a category four. Global warming is an imbalance of nature.

The premise of global warming is that industrial growth coupled with non-structured methods we as humans use to sustain ourselves has created a situation where our planet is getting hotter by the minute. We have seemingly negatively effected our environment by a cycle of harmful processes that now seem to be feeding upon themselves to exponentially increase the damage to our ecosystem.

Causes of Global Warming

Let us start our examination of Global warming with a study of its causes. Global warming is an overall state of existence that is the cumulative effect of hundreds of environmental factors. All of these join together in both a linear and random model to show global warming as a chain of events.

Most modern attention to the problem of global warming began with discussion of depletion of the Earth's Ozone layer. Ozone (O_3) is a molecular form of Oxygen. The Ozone layer is a relatively thin strata of these molecules set in the lower portion of the Earth's stratosphere.

Depletion of the Earth's Ozone layer has resulted in a large increase in Ultra Violet Radiation reaching the surface of the earth. Does this increase in UV rays equate to global warming? Not really. In fact most scientific opinion is that depletion of the Ozone layer results in cooling of both the stratosphere and troposphere.

So why mention depletion of the Ozone layer as regards to global warming? Because it represents a needed balance between harmful radiation being allowed to reach the earth's surface and our desire to stem the rapid increase in our air and water temperature. Remember, we are viewing global warming as a chain of events.

CAUSE OF GLOBAL WARMING

The primary cause of global warming is Carbon Dioxide emissions. CO_2 is being pumped into our atmosphere at an insane pace; 8 billion tons of CO_2 entered the air last year. Of course some of this is due to natural activity such as volcanic eruptions and people breathing.

But the Earth is equipped to easily absorb those into the normal regenerative process. No, the beginning of global warming was caused by fossil fuels being burned and emitting plenty of CO_2.

Fig. Major cause of Global Warming

Currently in the world 40% of all CO_2 emissions are caused by power plants. These are burning coal, natural gas and diesel fuel. Some power plants burn garbage. Some burn methane made from garbage. And discounting those super green electrical generating plants designed to issue negligible pollutants, all of our power plants let loose into the atmosphere CO_2.

33% of all the CO_2 sent forth is the product of cars and trucks. Internal combustion engines burning fossil fuels...gasoline and diesel spew forth a retching amount of CO_2.

3. 5% of all CO_2 emissions are released from aircraft traveling our friendly skies. Unfortunately, jets and other aircraft deliver their payload of pollutants directly into the troposphere.

12% of all CO_2 released into the atmosphere is related to buildings. This figure varies from one source to the next. Some place the per centage of emissions from buildings as high as 33%. What most of these figures do not address is the actual cause of the CO_2 emissions. In newly constructed buildings, production of materials used in building and energy used during construction are sited as the cause of carbon dioxide emissions. In existing buildings the CO_2 created by the energy upkeep of the building is the root of the emissions quotient.

The general comparison is that buildings consume energy in the way that cars burn fuel. But the pollutants created in providing power for heating, air-conditioning, lights and other usage in buildings has already been factored. Honestly this double billing accounting is more the product of auto manufacturers looking to point the blame for global warming away from gas guzzling cars.

The point to remember is that 98% of all CO_2 emissions are related to energy production and 80% of these emissions become greenhouse gases. Which now mentioned allows us to follow our chain of event's leading to global warming into the next most defined cause... Methane gas.

Methane is released into the atmosphere from a dozen major sources. These include natural and man made emissions. Natural release of Methane is primarily from wetlands, (including agriculture) termites, the ocean, and hydrates. Non-organic releases are based from, landfills, livestock, waste treatment, and biomass burning.

Almost all of this is offset by the Earth's ability to absorb around 97% of the methane released into the air. But that remaining 3% is a serious problem. The molecular structure of Methane makes it 20 times as powerful a Greenhouse gas than CO_2. So while there is a great deal less Methane to contend with than CO_2, it is still the second largest link in the global warming events chain.

Not every Greenhouse gas is as obvious a villain as Methane. The next most potent problem is simple H_2O water. How can water be a cause of global warming? Our atmosphere contains a set parameter of water as vapour. This vapour absorbs and radiates heat as does every molecule in the air.

But when the lower atmosphere (troposphere) has excess water vapour that gaseous H_2O is a potent greenhouse gas. Another of the more commonplace greenhouse gases is Nitrous Oxide. NO_2 can make your car go faster, or make you relax at the dentist. It has quite few beneficial uses. But as a greenhouse gas all it manages to accomplish is to be one more ingredient in out atmospheric soup. Cars using catalytic converters, fertilizer plants, manufacture of nylon, and nitric acid as well as being produced naturally in our oceans and rain forests, produce Nitrous Oxide.

All of the above plus quite a few other greenhouse gases comprise the foundation of global warming. As above and in all discussion of global warming they are cumulatively referred to as greenhouse gases. To understand the importance of these as the start and endpoint of global warming we must digress into a brief explanation of the greenhouse effect.

EFFECTS OF GLOBAL WARMING

The effects of global warming are of concern both for the environment and human life. Scenarios studied by the Intergovernmental Panel on Climate Change (IPCC) predict that global warming will continue and get worse much faster than was expected even in their last report. Research by NOAA indicate that the effects of global warming are already largely irreversible. The IPCC reports attribute many specific natural phenomena to human causes. The expected long range effects of recent climate change may already be observed.

Rising sea levels, glacier retreat, Arctic shrinkage, and altered patterns of agriculture are cited as direct consequences of human activities. Predictions for secondary and regional effects include extreme weather events, an expansion of tropical diseases, changes in the timing of seasonal patterns in ecosystems, and drastic economic impact. Concerns have led to political activism advocating proposals to mitigate, or adapt to it.

Climate changes characterized as global warming are leading to large-scale irreversible effects at continental and global scales. The likelihood and magnitude of the effects are observed and predicted to be increasing and accelerating.

Many consequences of global warming once controversial or thought to be unlikely are now being observed. Arctic shrinkage and Arctic methane release, alongside large reductions in the Greenland and West Antarctic Ice Sheets, accelerated global warming due to carbon cycle feedbacks in the terrestrial biosphere, and releases of terrestrial carbon from permafrost regions and methane from hydrates in coastal sediments are accelerating.

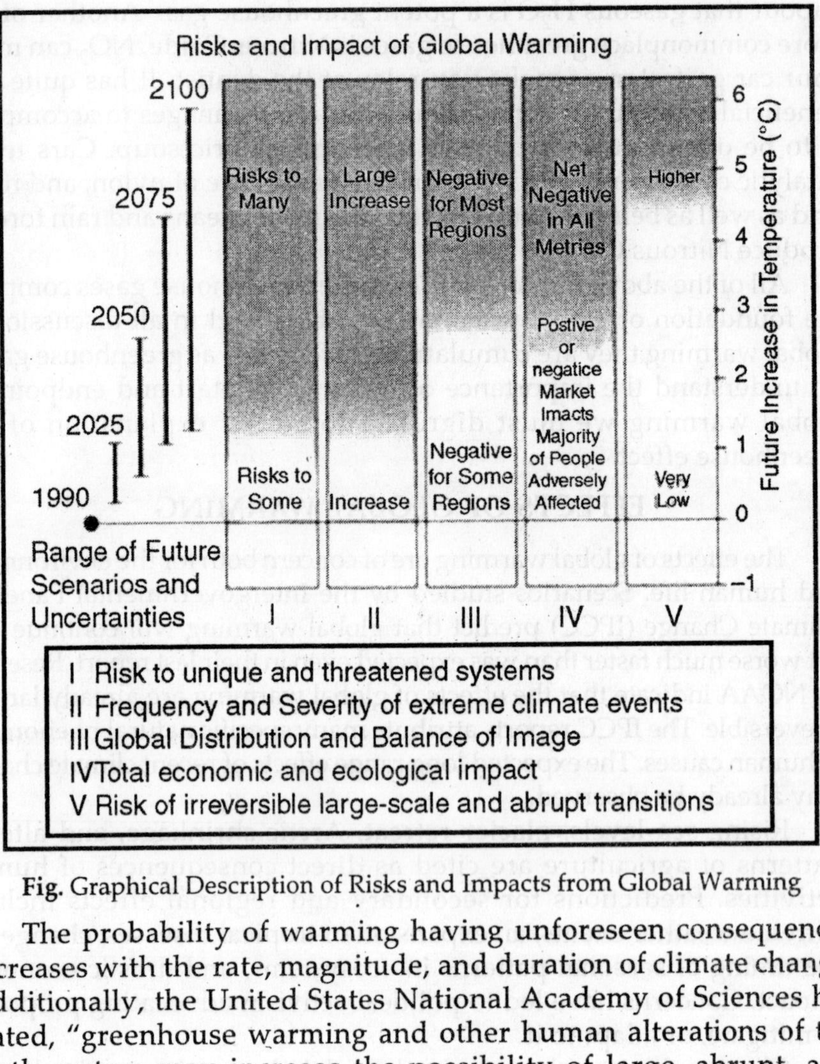

Fig. Graphical Description of Risks and Impacts from Global Warming

The probability of warming having unforeseen consequences increases with the rate, magnitude, and duration of climate change. Additionally, the United States National Academy of Sciences has stated, "greenhouse warming and other human alterations of the earth system may increase the possibility of large, abrupt, and unwelcome regional or global climatic events.... Future abrupt changes cannot be predicted with confidence, and climate surprises are to be expected."

The IPCC reports that the effects of global warming will be mixed across regions. For smaller values of warming (of up to 3°C, or about 5°F), changes are expected to produce net benefits in some regions and for some activities, and net costs for others. Greater warming may produce net costs (or to reduce the benefits from smaller warming) in all regions. Developing countries are vulnerable to reduced economic growth as a result of warming.

Most of the consequences of global warming would result from physical changes: sea level rise, higher local temperatures, and changes in rainfall patterns. Sea level is expected to rise 18 to 59 cm (7. 1 to 23. 2 inches) by the end of the 21st century, not including the unknown contribution from non-linear changes to large ice sheets.

It has also been proposed that the melting in the Arctic may bring fresh water to the North Atlantic to disrupt the Gulf Stream, which may cause a destabilisation or shutdown of the Thermohaline circulation.

PHYSICAL IMPACTS

Effects on Weather

Increasing temperature is likely to lead to increasing precipitationbut the effects on storms are less clear. Extratropical storms partly depend on the temperature gradient, which is predicted to weaken in the northern hemisphere as the polar region warms more than the rest of the hemisphere.

Extreme Weather

Storm strength leading to extreme weather is increasing, such as the power dissipation index of hurricane intensity. Kerry Emanuel writes that hurricane power dissipation is highly correlated with temperature, reflecting global warming. However, a further study by Emanuel using current model output concluded that the increase in power dissipation in recent decades cannot be completely attributed to global warming.

Hurricane modeling has produced similar results, finding that hurricanes, simulated under warmer, high-CO_2 conditions, are more intense, however, hurricane frequency will be reduced. Worldwide, the proportion of hurricanes reaching categories 4 or 5 – with wind speeds above 56 metres per second – has risen from 20% in the 1970s to 35% in the 1990s. Precipitation hitting the US from hurricanes has increased by 7% over the twentieth century.

The extent to which this is due to global warming as opposed to the Atlantic Multidecadal Oscillation is unclear. Some studies have found that the increase in sea surface temperature may be offset by an increase in wind shear, leading to little or no change in hurricane activity.

Hoyos *et al.* have linked the increasing trend in number of category 4 and 5 hurricanes for the period 1970-2004 directly to the trend in sea surface temperatures.

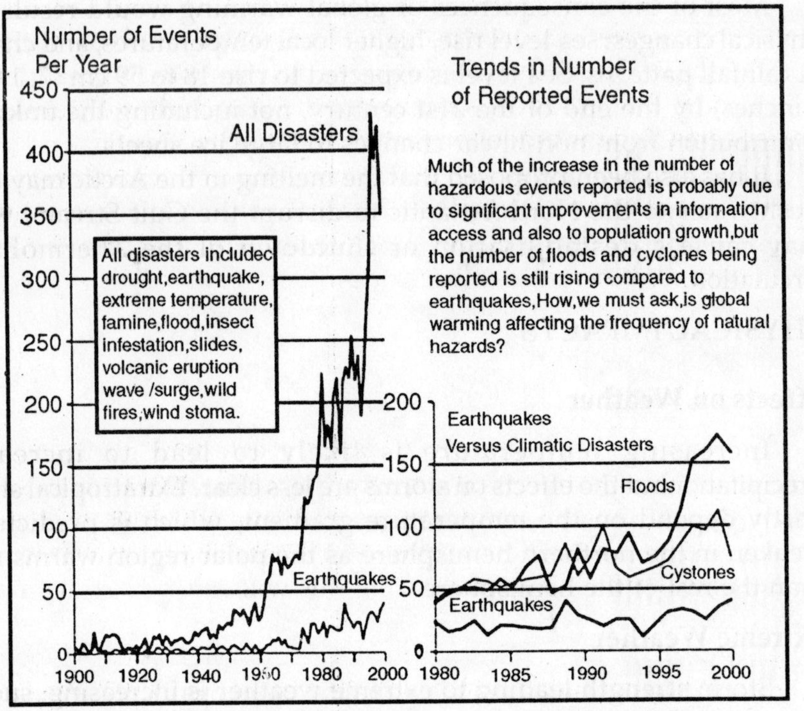

Fig. Global Warming may be Responsible in Part for some Trends in Natural Disasters Such as Extreme Weather

Increases in catastrophes resulting from extreme weather are mainly caused by increasing population densities, and anticipated future increases are similarly dominated by societal change rather than climate change.

The World Meteorological Organization explains that "though there is evidence both for and against the existence of a detectable anthropogenic signal in the tropical cyclone climate record to date, no firm conclusion can be made on this point. " They also clarified that "no individual tropical cyclone can be directly attributed to climate change. "

Thomas Knutson and Robert E. Tuleya of NOAA stated in 2004 that warming induced by greenhouse gas may lead to increasing occurrence of highly destructive category-5 storms. In 2008, Knutson *et al.* found that Atlantic hurricane and tropical storm frequencies could reduce under future greenhouse-gas-induced warming.

Vecchi and Soden find that wind shear, the increase of which acts to inhibit tropical cyclones, also changes in model-projections of global warming. There are projected increases of wind shear in the tropical Atlantic and East Pacific associated with the deceleration of the Walker circulation, as well as decreases of wind shear in the western and central Pacific. The study does not make claims about the net effect on Atlantic and East Pacific hurricanes of the warming and moistening atmospheres, and the model-projected increases in Atlantic wind shear.

A substantially higher risk of extreme weather does not necessarily mean a noticeably greater risk of slightly-above-average weather. However, the evidence is clear that severe weather and moderate rainfall are also increasing. Increases in temperature are expected to produce more intense convection over land and a higher frequency of the most severe storms.

INCREASED EVAPORATION

Over the course of the 20th century, evaporation rates have reduced worldwide; this is thought by many to be explained by global dimming. As the climate grows warmer and the causes of global dimming are reduced, evaporation will increase due to warmer oceans. Because the world is a closed system this will cause heavier rainfall, with more erosion.

This erosion, in turn, can in vulnerable tropical areas (especially in Africa) lead to desertification. On the other hand, in other areas, increased rainfall lead to growth of forests in dry desert areas.

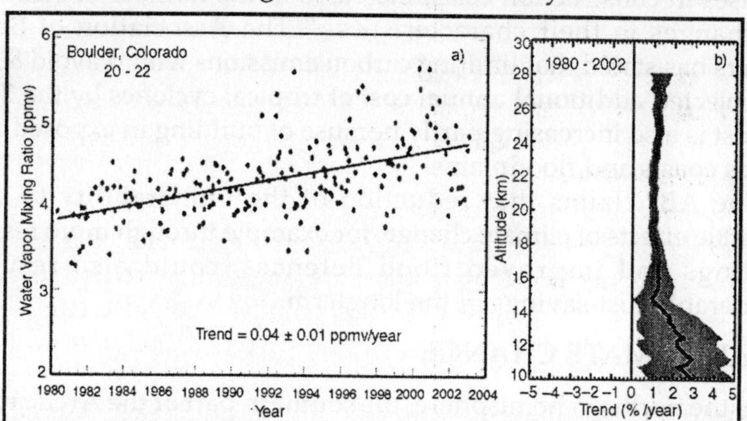

Fig. Increasing Water Vapour at Boulder, Colorado.

Scientists have found evidence that increased evaporation could result in more extreme weather as global warming progresses. The IPCC Third Annual Report says: ". . . global average water vapour concentration and precipitation are projected to increase during the 21st century.

By the second half of the 21st century, it is likely that precipitation will have increased over northern mid- to high latitudes and Antarctica in winter. At low latitudes there are both regional increases and decreases over land areas. Larger year to year variations in precipitation are very likely over most areas where an increase in mean precipitation is projected. "

COST OF MORE EXTREME WEATHER

As the World Meteorological Organization explains, "recent increase in societal impact from tropical cyclones has largely been caused by rising concentrations of population and infrastructure in coastal regions. " Pielke *et al.* (2008) normalized mainland U. S. hurricane damage from 1900–2005 to 2005 values and found no remaining trend of increasing absolute damage.

The 1970s and 1980s were notable because of the extremely low amounts of damage compared to other decades. The decade 1996–2005 has the second most damage among the past 11 decades, with only the decade 1926–1935 surpassing its costs. The most damaging single storm is the 1926 Miami hurricane, with $157 billion of normalized damage.

The American *Insurance Journal* predicted that "catastrophe losses should be expected to double roughly every 10 years because of increases in construction costs, increases in the number of structures and changes in their characteristics. " The Association of British Insurers has stated that limiting carbon emissions would avoid 80% of the projected additional annual cost of tropical cyclones by the 2080s. The cost is also increasing partly because of building in exposed areas such as coasts and floodplains.

The ABI claims that reduction of the vulnerability to some inevitable effects of climate change, for example through more resilient buildings and improved flood defences, could also result in considerable cost-savings in the longterm.

LOCAL CLIMATE CHANGE

In the northern hemisphere, the southern part of the Arctic region (home to 4,000,000 people) has experienced a temperature rise of 1 °C to 3 °C (1. 8 °F to 5. 4 °F) over the last 50 years. Canada, Alaška and Russia are experiencing initial melting of permafrost.

This may disrupt ecosystems and by increasing bacterial activity in the soil lead to these areas becoming carbon sources instead of carbon sinks. A study of changes to eastern Siberia's permafrost suggests that it is gradually disappearing in the southern regions, leading to the loss of nearly 11% of Siberia's nearly 11,000 lakes since 1971. At the same time, western Siberia is at the initial stage where melting permafrost is creating new lakes, which will eventually start disappearing as in the east. Furthermore, permafrost melting will eventually cause methane release from melting permafrost peat bogs.

Hurricanes were thought to be an entirely North Atlantic phenomenon. In late March 2004, the first Atlantic cyclone to form south of the equator hit Brazil with 40 m/s (144 km/h) winds, although some Brazilian meteorologists deny that it was a hurricane. Monitoring systems may have to be extended 1,600 km (1,000 miles) further south. There is no agreement as to whether this hurricane is linked to climate change, but one climate model exhibits increased tropical cyclone genesis in the South Atlantic under global warming by the end of the 21st century.

GLACIER RETREAT AND DISAPPEARANCE

In historic times, glaciers grew during a cool period from about 1550 to 1850 known as the Little Ice Age. Subsequently, until about 1940, glaciers around the world retreated as the climate warmed. Glacier retreat declined and reversed in many cases from 1950 to 1980 as a slight global cooling occurred.

Since 1980, glacier retreat has become increasingly rapid and ubiquitous, and has threatened the existence of many of the glaciers of the world. This process has increased markedly since 1995.

Excluding the ice caps and ice sheets of the Arctic and Antarctic, the total surface area of glaciers worldwide has decreased by 50% since the end of the 19th century. Currently glacier retreat rates and mass balance losses have been increasing in the Andes, Alps, Pyrenees, Himalayas, Rocky Mountains and North Cascades.

The loss of glaciers not only directly causes landslides, flash floods and glacial lake overflow, but also increases annual variation in water flows in rivers. Glacier runoff declines in the summer as glaciers decrease in size, this decline is already observable in several regions. Glaciers retain water on mountains in high precipitation years, since the snow cover accumulating on glaciers protects the ice from melting. In warmer and drier years, glaciers offset the lower precipitation amounts with a higher meltwater input.

Of particular importance are the Hindu Kush and Himalayan glacial melts that comprise the principal dry-season water source of many of the major rivers of the Central, South, East and Southeast Asian mainland. Increased melting would cause greater flow for several decades, after which "some areas of the most populated regions on Earth are likely to 'run out of water'" as source glaciers are depleted.

According to a UN climate report, the Himalayan glaciers that are the sources of Asia's biggest rivers—Ganges, Indus, Brahmaputra, Yangtze, Mekong, Salween and Yellow—could disappear by 2035 as temperatures rise. Approximately 2. 4 billion people live in the drainage basin of the Himalayan rivers.

India, China, Pakistan, Bangladesh, Nepal and Myanmar could experience floods followed by droughts in coming decades. In India alone, the Ganges provides water for drinking and farming for more than 500 million people. It has to be acknowledged, however, that increased seasonal runoff of Himalayan glaciers led to increased agricultural production in northern India throughout the 20th century.

The recession of mountain glaciers, notably in Western North America, Franz-Josef Land, Asia, the Alps, the Pyrenees, Indonesia and Africa, and tropical and sub-tropical regions of South America, has been used to provide qualitative support to the rise in global temperatures since the late 19th century.

Many glaciers are being lost to melting further raising concerns about future local water resources in these glaciated areas. In Western North America the 47 North Cascade glaciers observed all are retreating.

Despite their proximity and importance to human populations, the mountain and valley glaciers of temperate latitudes amount to a small fraction of glacial ice on the earth. About 99% is in the great ice sheets of polar and subpolar Antarctica and Greenland.

These continuous continental-scale ice sheets, 3 kilometres (1. 9 mi) or more in thickness, cap the polar and subpolar land masses. Like rivers flowing from an enormous lake, numerous outlet glaciers transport ice from the margins of the ice sheet to the ocean.

Glacier retreat has been observed in these outlet glaciers, resulting in an increase of the ice flow rate. In Greenland the period since the year 2000 has brought retreat to several very large glaciers that had long been stable. Three glaciers that have been researched, Helheim, Jakobshavn Isbræ and Kangerdlugssuaq Glaciers, jointly drain more than 16% of the Greenland Ice Sheet. Satellite images and aerial photographs from the 1950s and 1970s show that the front of the glacier had remained in the same place for decades.

But in 2001 it began retreating rapidly, retreating 7. 2 km (4. 5 mi) between 2001 and 2005. It has also accelerated from 20 m (66 ft)/day to 32 m (100 ft)/day. Jakobshavn Isbræ in western Greenland had been moving at speeds of over 24 m (79 ft)/day with a stable terminus since at least 1950. The glacier's ice tongue began to break apart in 2000, leading to almost complete disintegration in 2003, while the retreat rate doubled to over 30 m (98 ft)/day.

OCEANS

The role of the oceans in global warming is a complex one. The oceans serve as a sink for carbon dioxide, taking up much that would otherwise remain in the atmosphere, but increased levels of CO_2 have led to ocean acidification. Furthermore, as the temperature of the oceans increases, they become less able to absorb excess CO_2.

Global warming is projected to have a number of effects on the oceans. Ongoing effects include rising sea levels due to thermal expansion and melting of glaciers and ice sheets, and warming of the ocean surface, leading to increased temperature stratification. Other possible effects include large-scale changes in ocean circulation.

Sea Level Rise

With increasing average global temperature, the water in the oceans expands in volume, and additional water enters them which had previously been locked up on land in glaciers, for example, the Greenland and the Antarctic ice sheets. For most glaciers worldwide, an average volume loss of 60% until 2050 is predicted.

Meanwhile, the estimated total ice melting rate over Greenland is 239 ± 23 cubic kilometres (57 ± 5. 5 cu mi) per year, mostly from East Greenland. The Antarctic ice sheet, however, is expected to grow during the 21st century because of increased precipitation.

Under the IPCC Special Report on Emission Scenario (SRES) A1B, by the mid-2090s global sea level will reach 0. 22 to 0. 44 m (8. 7 to 17 in) above 1990 levels, and is currently rising at about 4 mm (0. 16 in) per year. Since 1900, the sea level has risen at an average of 1. 7 mm (0. 067 in) per year; since 1993, satellite altimetry from TOPEX/Poseidon indicates a rate of about 3 mm (0. 12 in) per year.

The sea level has risen more than 120 metres (390 ft) since the Last Glacial Maximum about 20,000 years ago. The bulk of that occurred before 7000 years ago. Global temperature declined after the Holocene Climatic Optimum, causing a sea level lowering of 0. 7 ± 0. 1 m (28 ± 3. 9 in) between 4000 and 2500 years before present.

From 3000 years ago to the start of the 19th century, sea level was almost constant, with only minor fluctuations. However, the Medieval Warm Period may have caused some sea level rise; evidence has been found in the Pacific Ocean for a rise to perhaps 0. 9 m (2 ft 11 in) above present level in 700 BP.

In a paper published in 2007, the climatologist James Hansen *et al.* claimed that ice at the poles does not melt in a gradual and linear fashion, but that another according to the geological record, the ice sheets can suddenly destabilize when a certain threshold is exceeded.

In this paper Hansen *et al.* state: Our concern that BAU GHG scenarios would cause large sealevel rise this century differs from estimates of IPCC (2001, 2007), which foresees little or no contribution to twentyfirst century sealevel rise from Greenland and Antarctica.

However, the IPCC analyses and projections do not well account for the nonlinear physics of wet ice sheet disintegration, ice streams and eroding ice shelves, nor are they consistent with the palaeoclimate evidence we have presented for the absence of discernible lag between ice sheet forcing and sealevel rise.

A paper published in 2008 by a group of researchers at the University of Wisconsin lead by Anders Carlson used the deglaciation of North America at 9000 years before present as an analogue to predict sea level rise of 1. 3 meters in the next century, which is also much higher than the IPCC predictions.

However, models of glacial flow in the smaller present-day ice sheets show that a probable maximum value for sea level rise in the next century is 80 centimeters, based on limitations on how quickly ice can flow below the equilibrium line altitude and to the sea.

TEMPERATURE RISE

From 1961 to 2003, the global ocean temperature has risen by 0. 10 °C from the surface to a depth of 700 m. There is variability both year-to-year and over longer time scales, with global ocean heat content observations showing high rates of warming for 1991 to 2003, but some cooling from 2003 to 2007.

The temperature of the Antarctic Southern Ocean rose by 0. 17 °C (0. 31 °F) between the 1950s and the 1980s, nearly twice the rate for the world's oceans as a whole. As well as having effects on ecosystems (e. g. by melting sea ice, affecting algae that grow on its underside), warming reduces the ocean's ability to absorb CO_2.

ACIDIFICATION

Ocean acidification is an effect of rising concentrations of CO_2 in the atmosphere, and is not a direct consequence of global warming. The oceans soak up much of the CO_2 produced by living organisms, either as dissolved gas, or in the skeletons of tiny marine creatures that fall to the bottom to become chalk or limestone.

Oceans currently absorb about one tonne of CO_2 per person per year. It is estimated that the oceans have absorbed around half of all CO_2 generated by human activities since 1800 (118 ± 19 petagrams of carbon from 1800 to 1994).

In water, CO_2 becomes a weak carbonic acid, and the increase in the greenhouse gas since the industrial revolution has already lowered the average pH (the laboratory measure of acidity) of seawater by 0. 1 units, to 8. 2. Predicted emissions could lower the pH by a further 0. 5 by 2100, to a level probably not seen for hundreds of millennia and, critically, at a rate of change probably 100 times greater than at any time over this period.

There are concerns that increasing acidification could have a particularly detrimental effect on corals (16% of the world's coral reefs have died from bleaching caused by warm water in 1998, which coincidentally was the warmest year ever recorded) and other marine organisms with calcium carbonate shells.

Shutdown of Thermohaline Circulation

There is some speculation that global warming could, via a shutdown or slowdown of the thermohaline circulation, trigger localized cooling in the North Atlantic and lead to cooling, or lesser warming, in that region. This would affect in particular areas like Scandinavia and Britain that are warmed by the North Atlantic drift.

The chances of this near-term collapse of the circulation are unclear; there is some evidence for the short-term stability of the Gulf Stream and possible weakening of the North Atlantic drift. However, the degree of weakening, and whether it will be sufficient to shut down the circulation, is under debate. As yet, no cooling has been found in northern Europe or nearby seas. Lenton et al. found that "simulations clearly pass a THC tipping point this century".

Methane Release from Melting Permafrost Peat Bogs

Western Siberia is the world's largest peat bog, a one million square kilometre region of permafrost peat bog that was formed 11,000 years ago at the end of the last ice age. The melting of its permafrost is likely to lead to the release, over decades, of large quantities of methane.

As much as 70,000 million tonnes of methane, an extremely effective greenhouse gas, might be released over the next few decades, creating an additional source of greenhouse gas emissions. Similar melting has been observed in eastern Siberia. Lawrence et al. (2008) suggest that a rapid melting of Arctic sea ice may start a feedback loop that rapidly melts Arctic permafrost, triggering further warming.

Methane Release from Hydrates

Methane clathrate, also called methane hydrate, is a form of water ice that contains a large amount of methane within its crystal structure. Extremely large deposits of methane clathrate have been found under sediments on the ocean floors of Earth. The sudden release of large amounts of natural gas from methane clathrate deposits, in a runaway greenhouse effect, has been hypothesized as a cause of past and possibly future climate changes. The release of this trapped methane is a potential major outcome of a rise in temperature; it is thought that this might increase the global temperature by an additional 5° in itself, as methane is much more powerful as a greenhouse gas than carbon dioxide.

The theory also predicts this will greatly affect available oxygen content of the atmosphere. This theory has been proposed to explain the most severe mass extinction event on earth known as the Permian-Triassic extinction event.

In 2008, a research expedition for the American Geophysical Union detected levels of methane up to 100 times above normal in the Siberian Arctic, likely being released by methane clathrates being released by holes in a frozen 'lid' of seabed permafrost, around the outfall of the Lena River and the area between the Laptev Sea and East Siberian Sea.

CARBON CYCLE FEEDBACKS

There have been predictions, and some evidence, that global warming might cause loss of carbon from terrestrial ecosystems, leading to an increase of atmospheric CO_2 levels. Several climate models indicate that global warming through the 21st century could be accelerated by the response of the terrestrial carbon cycle to such warming.

All 11 models in the C4MIP study found that a larger fraction of anthropogenic CO_2 will stay airborne if climate change is accounted for. By the end of the twenty-first century, this additional CO_2 varied between 20 and 200 ppm for the two extreme models, the majority of the models lying between 50 and 100 ppm. The higher CO_2 levels led to an additional climate warming ranging between 0. 1° and 1. 5 °C. However, there was still a large uncertainty on the magnitude of these sensitivities.

Eight models attributed most of the changes to the land, while three attributed it to the ocean. The strongest feedbacks in these cases are due to increased respiration of carbon from soils throughout the high latitude boreal forests of the Northern Hemisphere. One model in particular (HadCM3) indicates a secondary carbon cycle feedback due to the loss of much of the Amazon rainforest in response to significantly reduced precipitation over tropical South America. While models disagree on the strength of any terrestrial carbon cycle feedback, they each suggest any such feedback would accelerate global warming. Observations show that soils in England have been losing carbon at the rate of four million tonnes a year for the past 25 years according to a paper in Nature by Bellamy et al. in September 2005, who note that these results are unlikely to be explained by land use changes. Results such as this rely on a dense sampling network and thus are not available on a global scale.

Extrapolating to all of the United Kingdom, they estimate annual losses of 13 million tons per year. This is as much as the annual reductions in carbon dioxide emissions achieved by the UK under the Kyoto Treaty (12. 7 million tons of carbon per year).

It has also been suggested (by Chris Freeman) that the release of dissolved organic carbon (DOC) from peat bogs into water courses (from which it would in turn enter the atmosphere) constitutes a positive feedback for global warming.

The carbon currently stored in peatlands (390-455 gigatonnes, one-third of the total land-based carbon store) is over half the amount of carbon already in the atmosphere. DOC levels in water courses are observably rising; Freeman's hypothesis is that, not elevated temperatures, but elevated levels of atmospheric CO_2 are responsible, through stimulation of primary productivity.

Tree deaths are believed to be increasing as a result of climate change, which is a positive feedback effect. This contradicts the previously widely-held view that increased natural vegetation would lead to a negative-feedback effect.

FOREST FIRES

The IPCC Fourth Assessment Report predicts that many mid-latitude regions, such as Mediterranean Europe, will experience decreased rainfall and an increased risk of drought, which in turn would allow forest fires to occur on larger scale, and more regularly. This releases more stored carbon into the atmosphere than the carbon cycle can naturally re-absorb, as well as reducing the overall forest area on the planet, creating a positive feedback loop.

Part of that feedback loop is more rapid growth of replacement forests and a northward migration of forests as northern latitudes become more suitable climates for sustaining forests. There is a question of whether the burning of renewable fuels such as forests should be counted as contributing to global warming. Cook & Vizy also found that forest fires were likely in the Amazon rainforest, eventually resulting in a transition to Caatinga vegetation in the Eastern Amazon region.

RETREAT OF SEA ICE

The sea absorbs heat from the sun, while the ice largely reflects the sun rays back to space. Thus, retreating sea ice will allow the sun to warm the now exposed sea water, contributing to further warming. The mechanism is the same as when a black car heats up faster in sunlight than a white car.

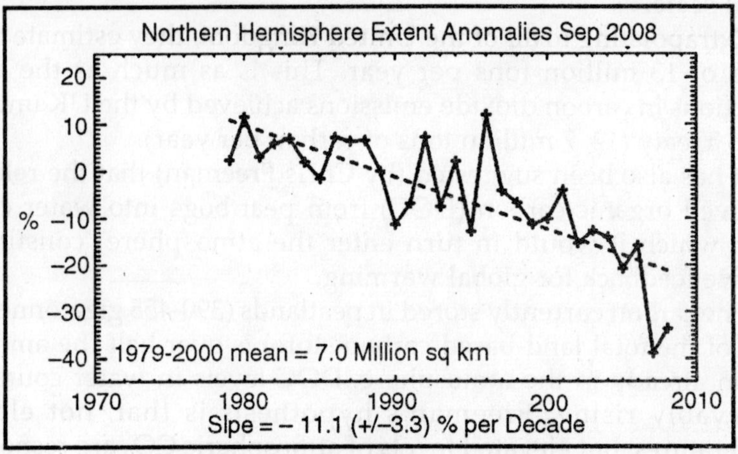

Fig. Northern Hemisphere Ice Trends

This albedo change is also the main reason why IPCC predict polar temperatures in the northern hemisphere to rise up to twice as much as those of the rest of the world. In September 2007, the Arctic sea ice area reached about half the size of the average summer minimum area between 1979 to 2000. Based on the accelerated loss, predictions are that by 2030 the Arctic could be ice-free part of the year.

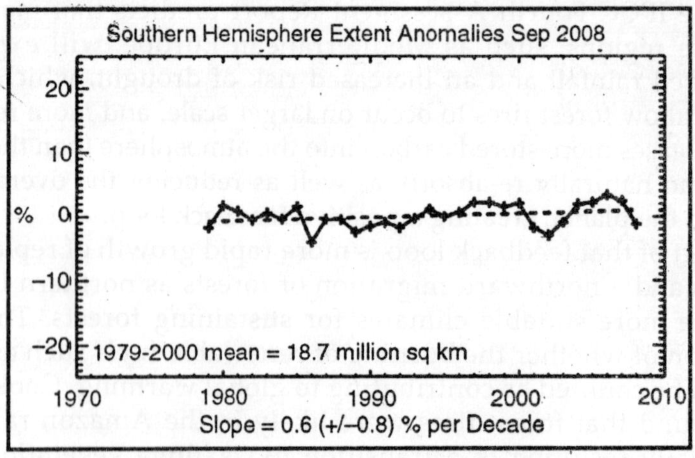

Fig. Southern Hemisphere Ice Trends

Also in September 2007, Arctic sea ice retreated far enough for the Northwest Passage to become navigable to shipping for the first time in recorded history. The polar amplification of global warming is not predicted to occur in the southern hemisphere. The Antarctic sea ice reached its greatest extent on record since the beginning of observation in 1979, but the gain in ice in the south is exceeded by the loss in the north. The trend for global sea ice, northern hemisphere and southern hemisphere combined is clearly a decline.

EFFECT ON SULFUR AEROSOLS

Sulfur aerosols, especially stratospheric sulfur aerosols have a significant effect on climate. One source of such aerosols is the sulfur cycle, where plankton release gases such as DMS which eventually becomes oxidised to sulfur dioxide in the atmosphere. Disruption to the oceans as a result of ocean acidification or disruptions to the thermohaline circulation may result in disruption of the sulfur cycle, thus reducing its cooling effect on the planet through the creation of stratospheric sulfur aerosols.

Economic and Social

The IPCC Fourth Assessment Report contains a review of the literature on the economics of climate change:

Impacts of climate change are *very likely* to impose net annual costs, which will increase over time as global temperatures increase. Peer-reviewed estimates of the social cost of carbon [Net economic costs of damages from climate change aggregated across the globe and discounted to the specified year] in 2005 average US$12 per tonne of CO_2, but the range from 100 estimates is large (-$3 to $95/t$CO_2$).

This is due in large part to differences in assumptions regarding climate sensitivity, response lags, the treatment of risk and equity, economic and non-economic impacts, the inclusion of potentially catastrophic losses and discount rates. Aggregate estimates of costs mask significant differences in impacts across sectors, regions and populations and *very likely* underestimate damage costs because they cannot include many non-quantifiable impacts.

In 2006, Nicholas Stern, the former Chief Economist and Senior Vice-President of the World Bank, presented the *Stern Review on the Economics of Climate Change*. Stern recommended that one per cent of global GDP be invested to mitigate the effects of climate change, and that failure to do so could risk a recession worth up to twenty per cent of global GDP. Stern's report suggests that climate change threatens to be the greatest and widest-ranging market failure ever seen. The report has had significant political effects: Australia reported two days after the report was released that they would allott AU$60 million to projects to help cut greenhouse gas emissions. More recently, Stern has stated that his original report underestimated the damages of climate change.

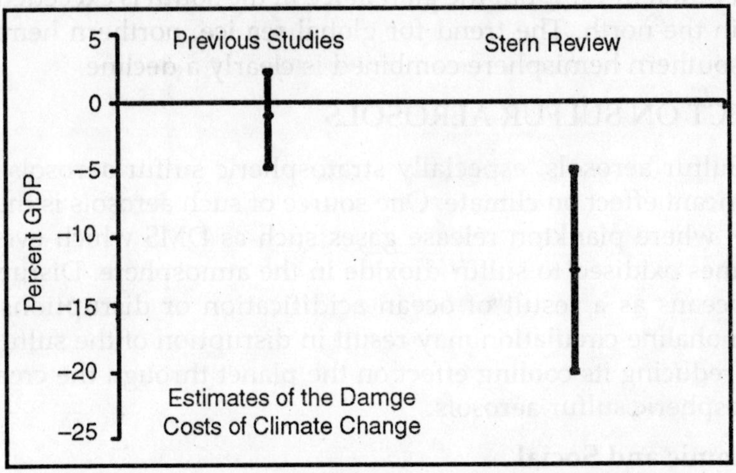

Fig. As Recent Estimates of the Rate of Global Warming have Increased, so have the Financial Estimates of the Damage Costs

The *Stern Review* has been criticized by some economists, saying that Stern did not consider costs past 2200, that he used an incorrect discount rate in his calculations, and that stopping or significantly slowing climate change will require deep emission cuts everywhere. Other economists have supported Stern's approach, or argued that Stern's estimates are reasonable, even if the method by which he reached them is open to criticism. .

In a 2004 comment on the economic effect of global warming in Copenhagen Consensus, Professor Robert O. Mendelsohn of Yale School of Forestry and Environmental Studies, stated that:

A series of studies on the impacts of climate change have systematically shown that the older literature overestimated climate damages by failing to allow for adaptation and for climate benefits. These new studies imply that impacts depend heavily upon initial temperatures (latitude).

Countries in the polar region are likely to receive large benefits from warming, countries in the mid-latitudes will at first benefit and only begin to be harmed if temperatures rise above 2. 5 °C (4. 50 °F). Only countries in the tropical and subtropical regions are likely to be harmed immediately by warming and be subject to the magnitudes of ɿpacts first thought likely. Summing these regional impacts across ɩe globe implies that warming benefits and damages will likely offset ach other until warming passes 2. 5°C and even then it will be far smaller on net than originally thought.

INSURANCE

An industry very directly affected by the risks is the insurance industry. According to a 2005 report from the Association of British Insurers, limiting carbon emissions could avoid 80% of the projected additional annual cost of tropical cyclones by the 2080s. A June 2004 report by the Association of British Insurers declared "Climate change is not a remote issue for future generations to deal with.

It is, in various forms, here already, impacting on insurers' businesses now. " It noted that weather risks for households and property were already increasing by 2-4 % per year due to changing weather, and that claims for storm and flood damages in the UK had doubled to over £6 billion over the period 1998–2003, compared to the previous five years. The results are rising insurance premiums, and the risk that in some areas flood insurance will become unaffordable for some.

Financial institutions, including the world's two largest insurance companies, Munich Re and Swiss Re, warned in a 2002 study that "the increasing frequency of severe climatic events, coupled with social trends" could cost almost US$ 150 billion each year in the next decade. These costs would, through increased costs related to insurance and disaster relief, burden customers, taxpayers, and industry alike.

In the United States, insurance losses have also greatly increased. According to Choi and Fisher (2003) each 1% increase in annual precipitation could enlarge catastrophe loss by as much as 2. 8%. Gross increases are mostly attributed to increased population and property values in vulnerable coastal areas, though there was also an increase in frequency of weather-related events like heavy rainfalls since the 1950s.

TRANSPORT

Roads, airport runways, railway lines and pipelines, (including oil pipelines, sewers, water mains etc) may require increased maintenance and renewal as they become subject to greater temperature variation. Regions already adversely affected include areas of permafrost, which are subject to high levels of subsidence, resulting in buckling roads, sunken foundations, and severely cracked runways.

EFFECTS ON AGRICULTURE

For some time it was hoped that a positive effect of global warming would be increased agricultural yields, because of the role of carbon dioxide in photosynthesis, especially in preventing photorespiration, which is responsible for significant destruction of several crops.

In Iceland, rising temperatures have made possible the widespread sowing of barley, which was untenable twenty years ago. Some of the warming is due to a local (possibly temporary) effect via ocean currents from the Caribbean, which has also affected fish stocks.

There is evidence to suggest some areas like the US and Siberia will benefit from global warming. However, large-scale experiments have shown that "Rising atmospheric temperatures, longer droughts and side-effects of both, such as higher levels of ground-level ozone gas, are likely to bring about a substantial reduction in crop yields in the coming decades".

Moreover, the region likely to be worst affected is Africa, both because its geography makes it particularly vulnerable, and because seventy per cent of the population rely on rain-fed agriculture for their livelihoods. Tanzania's official report on climate change suggests that the areas that usually get two rainfalls in the year will probably get more, and those that get only one rainy season will get far less. The net result is expected to be that 33% less maize—the country's staple crop—will be grown.

Climate change may be one of the causes of the Darfur conflict. The combination of decades of drought, desertification and overpopulation are among the causes of the conflict, because the Arab Baggara nomads searching for water have to take their livestock further south, to land mainly occupied by farming peoples.

The scale of historical climate change, as recorded in Northern Darfur, is almost unprecedented: the reduction in rainfall has turned millions of hectares of already marginal semi-desert grazing land into desert. The impact of climate change is considered to be directly related to the conflict in the region, as desertification has added significantly to the stress on the livelihoods of pastoralist societies, forcing them to move south to find pasture

FLOOD DEFENCE

For historical reasons to do with trade, many of the world's largest and most prosperous cities are on the coast, and the cost of building better coastal defenses (due to the rising sea level) is likely to be considerable.

Some countries will be more affected than others—low-lying

countries such as Bangladesh and the Netherlands would be worst hit by any sea level rise, in terms of floods or the cost of preventing them. Still, in 180 of 192 littoral countries worldwide, coastal protection will cost less than 0. 1% of the country's gross domestic product.

In developing countries, the poorest often live on flood plains, because it is the only available space, or fertile agricultural land. These settlements often lack infrastructure such as dykes and early warning systems. Poorer communities also tend to lack the insurance, savings or access to credit needed to recover from disasters.

MIGRATION

Some Pacific Ocean island nations, such as Tuvalu, are concerned about the possibility of an eventual evacuation, as flood defence may become economically unviable for them. Tuvalu already has an ad hoc agreement with New Zealand to allow phased relocation.

In the 1990s a variety of estimates placed the number of environmental refugees at around 25 million. (Environmental refugees are not included in the official definition of refugees, which only includes migrants fleeing persecution.)

The Intergovernmental Panel on Climate Change (IPCC), which advises the world's governments under the auspices of the UN, estimated that 150 million environmental refugees will exist in the year 2050, due mainly to the effects of coastal flooding, shoreline erosion and agricultural disruption (150 million means 1. 5% of 2050's predicted 10 billion world population). Northwest Passage

Melting Arctic ice may open the Northwest Passage in summer, which would cut 5,000 nautical miles (9,000 km) from shipping routes between Europe and Asia. This would be of particular benefit for supertankers which are too big to fit through the Panama Canal and currently have to go around the tip of South America. According to the Canadian Ice Service, the amount of ice in Canada's eastern Arctic Archipelago decreased by 15% between 1969 and 2004.

In September 2007, the Arctic Ice Cap retreated far enough for the Northwest Passage to become navigable to shipping for the first time in recorded history.

In August, 2008, melting sea ice simultaneously opened up the Northwest Passage and the Northern Sea Route, making it possible to sail around the Arctic ice cap. Scientists estimate that this hasn't happened in 125,000 years.

The Northwest Passage opened August 25, 2008, and the remaining tongue of ice blocking the Northern Sea Route dissolved a few days later. Because of Arctic shrinkage, the Beluga group of Bremen,

Germany, announced plans to send the first ship through the Northern Sea Route in 2009.

DEVELOPMENT

The combined effects of global warming may have particularly harsh effects on people and countries without the resources to mitigate those effects. This may slow economic development and poverty reduction, and make it harder to achieve the Millennium Development Goals.

In October 2004 the Working Group on Climate Change and Development, a coalition of development and environment NGOs, issued a report Up in Smoke on the effects of climate change on development. This report, and the July 2005 report Africa - Up in Smoke? predicted increased hunger and disease due to decreased rainfall and severe weather events, particularly in Africa. These are likely to have severe impacts on development for those affected.

ECOSYSTEMS

Unchecked global warming could affect most terrestrial ecoregions. Increasing global temperature means that ecosystems will change; some species are being forced out of their habitats (possibly to extinction) because of changing conditions, while others are flourishing.

Secondary effects of global warming, such as lessened snow cover, rising sea levels, and weather changes, may influence not only human activities but also the ecosystem. Studying the association between Earth climate and extinctions over the past 520 million years, scientists from University of York write, "The global temperatures predicted for the coming centuries may trigger a new 'mass extinction event', where over 50 per cent of animal and plant species would be wiped out."

Many of the species at risk are Arctic and Antarctic fauna such as polar bears and emperor penguins. In the Arctic, the waters of Hudson Bay are ice-free for three weeks longer than they were thirty years ago, affecting polar bears, which prefer to hunt on sea ice. .

Species that rely on cold weather conditions such as gyrfalcons, and snowy owls that prey on lemmings that use the cold winter to their advantage may be hit hard. Marine invertebrates enjoy peak growth at the temperatures they have adapted to, regardless of how cold these may be, and cold-blooded animals found at greater latitudes and altitudes generally grow faster to compensate for the short growing season.

Warmer-than-ideal conditions result in higher metabolism and consequent reductions in body size despite increased foraging, which in turn elevates the risk of predation. Indeed, even a slight increase in temperature during development impairs growth efficiency and survival rate in rainbow trout.

Rising temperatures are beginning to have a noticeable impact on birds, and butterflies have shifted their ranges northward by 200 km in Europe and North America. Plants lag behind, and larger animals' migration is slowed down by cities and roads. In Britain, spring butterflies are appearing an average of 6 days earlier than two decades ago.

A 2002 article in *Nature* surveyed the scientific literature to find recent changes in range or seasonal behaviour by plant and animal species. Of species showing recent change, 4 out of 5 shifted their ranges towards the poles or higher altitudes, creating "refugee species". Frogs were breeding, flowers blossoming and birds migrating an average 2. 3 days earlier each decade; butterflies, birds and plants moving towards the poles by 6. 1 km per decade.

A 2005 study concludes human activity is the cause of the temperature rise and resultant changing species behaviour, and links these effects with the predictions of climate models to provide validation for them. Scientists have observed that Antarctic hair grass is colonizing areas of Antarctica where previously their survival range was limited.

Mechanistic studies have documented extinctions due to recent climate change: McLaughlin *et al.* documented two populations of Bay checkerspot butterfly being threatened by precipitation change. Parmesan states, "Few studies have been conducted at a scale that encompasses an entire species" and McLaughlin *et al.* agreed "few mechanistic studies have linked extinctions to recent climate change. " Daniel Botkin and other authors in one study believe that projected rates of extinction are overestimated.

Many species of freshwater and saltwater plants and animals are dependent on glacier-fed waters to ensure a cold water habitat that they have adapted to. Some species of freshwater fish need cold water to survive and to reproduce, and this is especially true with Salmon and Cutthroat trout. Reduced glacier runoff can lead to insufficient stream flow to allow these species to thrive.

Ocean krill, a cornerstone species, prefer cold water and are the primary food source for aquatic mammals such as the Blue whale. Alterations to the ocean currents, due to increased freshwater inputs from glacier melt, and the potential alterations to thermohaline circulation of the worlds oceans, may affect existing fisheries upon which humans depend as well. The *white lemuroid possum*, only found in the mountain forests of northern Queensland, has been named as the first mammal species to be driven extinct by man-made global warming. The White Possum has not been seen in over three years. These possums cannot survive extended temperatures over 30 °C (86 °F), which occurred in 2005. A final expedition to uncover any surviving White Possums is scheduled for 2009.

FORESTS

Pine forests in British Columbia have been devastated by a pine beetle infestation, which has expanded unhindered since 1998 at least in part due to the lack of severe winters since that time; a few days of extreme cold kill most mountain pine beetles and have kept outbreaks in the past naturally contained.

The infestation, which (by November 2008) has killed about half of the province's lodgepole pines (33 million acres or 135,000 km) is an order of magnitude larger than any previously recorded outbreak and passed via unusually strong winds in 2007 over the continental divide to Alberta. An epidemic also started, be it at a lower rate, in 1999 in Colorado, Wyoming, and Montana. The United States forest service predicts that between 2011 and 2013 virtually all 5 million acres (20,000 km) of Colorado's lodgepole pine trees over five inches (127 mm) in diameter will be lost.

As the northern forests are a carbon sink, while dead forests are a major carbon source, the loss of such large areas of forest has a positive feedback on global warming. In the worst years, the carbon emission due to beetle infestation of forests in British Columbia alone approaches that of an average year of forest fires in all of Canada or five years worth of emissions from that country's transportation sources.

Besides the immediate ecological and economic impact, the huge dead forests provide a fire risk. Even many healthy forests appear to face an increased risk of forest fires because of warming climates. The 10-year average of boreal forest burned in North America, after several decades of around 10,000 km^2 (2. 5 million acres), has increased steadily since 1970 to more than 28,000 km^2 (7 million acres) annually. .

Though this change may be due in part to changes in forest management practices, in the western U. S. , since 1986, longer, warmer summers have resulted in a fourfold increase of major wildfires and a sixfold increase in the area of forest burned, compared to the period from 1970 to 1986. A similar increase in wildfire activity has been reported in Canada from 1920 to 1999.

Forest fires in Indonesia have dramatically increased since 1997 as well. These fires are often actively started to clear forest for agriculture. They can set fire to the large peat bogs in the region and the CO_2 released by these peat bog fires has been estimated, in an average year, to be 15% of the quantity of CO_2 produced by fossil fuel combustion.

MOUNTAINS

Mountains cover approximately 25 per cent of earth's surface and provide a home to more than one-tenth of global human population. Changes in global climate pose a number of potential risks to mountain habitats. Researchers expect that over time, climate change will affect mountain and lowland ecosystems, the frequency and intensity of forest fires, the diversity of wildlife, and the distribution of water.

Studies suggest that a warmer climate in the United States would cause lower-elevation habitats to expand into the higher alpine zone. Such a shift would encroach on the rare alpine meadows and other high-altitude habitats. High-elevation plants and animals have limited space available for new habitat as they move higher on the mountains in order to adapt to long-term changes in regional climate.

Changes in climate will also affect the depth of the mountains snowpacks and glaciers. Any changes in their seasonal melting can have powerful impacts on areas that rely on freshwater runoff from mountains. Rising temperature may cause snow to melt earlier and faster in the spring and shift the timing and distribution of runoff. These changes could affect the availability of freshwater for natural systems and human uses.

ECOLOGICAL PRODUCTIVITY

Increasing average temperature and carbon dioxide may have the effect of improving ecosystems' productivity. In photorespiration, carbon dioxide that oxygen can enter a plant's chloroplasts and take the place of carbon dioxide in the Calvin cycle.

This causes the sugars being made to be destroyed, suppressing growth. Higher carbon dioxide concentrations tend to reduce photorespiration. Satellite data shows that the productivity of the northern hemisphere has increased since 1982 (although attribution of this increase to a specific cause is difficult).

IPCC models predict that higher CO_2 concentrations would only spur growth of flora up to a point, because in many regions the limiting factors are water or nutrients, not temperature or CO_2; after that, greenhouse effects and warming would continue but there would be no compensatory increase in growth.

Research done by the Swiss Canopy Crane Project suggests that slow-growing trees only are stimulated in growth for a short period under higher CO_2 levels, while faster growing plants like liana benefit in the long term. In general, but especially in rain forests, this means that liana become the prevalent species; and because they decompose much faster than trees their carbon content is more quickly returned to the atmosphere. Slow growing trees incorporate atmospheric carbon for decades.

WATER SCARCITY

Sea level rise is projected to increase salt-water intrusion into groundwater in some regions, affecting drinking water and agriculture in coastal zones. Increased evaporation will reduce the effectiveness of reservoirs. Increased extreme weather means more water falls on hardened ground unable to absorb it, leading to flash floods instead of a replenishment of soil moisture or groundwater levels. In some areas, shrinking glaciers threaten the water supply.

The continued retreat of glaciers will have a number of different effects. In areas that are heavily dependent on water runoff from glaciers that melt during the warmer summer months, a continuation of the current retreat will eventually deplete the glacial ice and substantially reduce or eliminate runoff. A reduction in runoff will affect the ability to irrigate crops and will reduce summer stream flows necessary to keep dams and reservoirs replenished.

This situation is particularly acute for irrigation in South America, where numerous artificial lakes are filled almost exclusively by glacial melt. Central Asian countries have also been historically dependent on the seasonal glacier melt water for irrigation and drinking supplies. In Norway, the Alps, and the Pacific Northwest of North America, glacier runoff is important for hydropower. Higher temperatures will also increase the demand for water for the purposes of cooling and hydration.

In the Sahel, there has been an unusually wet period from 1950 until 1970, followed by extremely dry years from 1970 to 1990. From 1990 until 2004 rainfall returned to levels slightly below the 1898–1993 average, but year-to-year variability was high.

HEALTH

Direct Effects of Temperature Rise

The most direct effect of climate change on humans might be the impacts of hotter temperatures themselves. Extreme high temperatures increase the number of people who die on a given day for many reasons: people with heart problems are vulnerable because one's cardiovascular system must work harder to keep the body cool during hot weather, heat exhaustion, and some respiratory problems increase.

Global warming could mean more cardiovascular diseases, doctors warn. Higher air temperature also increase the concentration of ozone at ground level. In the lower atmosphere, ozone is a harmful pollutant. It damages lung tissues and causes problems for people with asthma and other lung diseases.

Rising temperatures have two opposing direct effects on mortality: higher temperatures in winter reduce deaths from cold; higher temperatures in summer increase heat-related deaths. The net local impact of these two direct effects depends on the current climate in a particular area.

Palutikof *et al.* (1996) calculate that in England and Wales for a 1 °C temperature rise the reduced deaths from cold outweigh the increased deaths from heat, resulting in a reduction in annual average mortality of 7000, while Keatinge *et al.* (2000) "suggest that any increases in mortality due to increased temperatures would be outweighed by much larger short term declines in cold related mortalities."

Cold-related deaths are far more numerous than heat-related deaths in the United States, Europe, and almost all countries outside the tropics. During 1979–1999, a total of 3,829 deaths in the United States were associated with excessive heat due to weather conditions, while in that same period a total of 13,970 deaths were attributed to hypothermia. In Europe, mean annual heat related mortalities are 304 in North Finland, 445 in Athens, and 40 in London, while cold related mortalities are 2457, 2533, and 3129 respectively.

According to Keatinge *et al.* (2000), "populations in Europe have adjusted successfully to mean summer temperatures ranging from 13. 5°C to 24. 1°C, and can be expected to adjust to global warming predicted for the next half century with little sustained increase in heat related mortality."

A government report shows decreased mortality due to recent warming and predicts increased mortality due to future warming in the United Kingdom. The European heat wave of 2003 killed 22,000–35,000 people, based on normal mortality rates. Peter A. Stott from the Hadley Centre for Climate Prediction and Research estimated with 90% confidence that past human influence on climate was responsible for at least half the risk of the 2003 European summer heat-wave.

SPREAD OF DISEASE

Global warming may extend the favourable zones for vectors conveying infectious disease such as dengue fever, West Nile Virus, and malaria. In poorer countries, this may simply lead to higher incidence of such diseases. In richer countries, where such diseases have been eliminated or kept in check by vaccination, draining swamps and using pesticides, the consequences may be felt more in economic than health terms.

The World Health Organisation (WHO) says global warming could lead to a major increase in insect-borne diseases in Britain and Europe, as northern Europe becomes warmer, ticks—which carry encephalitis and lyme disease—and sandflies—which carry visceral leishmaniasis—are likely to move in. However, malaria has always been a common threat in European past, with the last epidemic occurring in the Netherlands during the 1950s.

In the United States, Malaria has been endemic in as much as 36 states (including Washington, North Dakota, Michigan and New York) until the 1940s. By 1949, the country was declared free of malaria as a significant public health problem, after more than 4,650,000 house DDT spray applications had been made.

The World Health Organisation estimates 150,000 deaths annually "as a result of climate change", of which half in the Asia-Pacific region. In April 2008, it reported that, as a result of increased temperatures, the number of malaria infections is expected to increase in the highland areas of Papua New Guinea.

CHILDREN

In 2007, the American Academy of Pediatrics issued the policy statement *Global Climate Change and Children's Health*:

Anticipated direct health consequences of climate change include injury and death from extreme weather events and natural disasters, increases in climate-sensitive infectious diseases, increases in air pollution–related illness, and more heat-related, potentially fatal, illness. Within all of these categories, children have increased vulnerability compared with other groups.

On 2008-04-29, a UNICEF UK Report found that global warming is already reducing the quality of the world's most vulnerable children's lives and making it more difficult to meet the UN Millennium Development Goals. Global warming will reduce access to clean water and food supplies, particularly in Africa and Asia. Disasters, violence and disease are expected to be more frequent and intense, making the future of the world's poorest children more bleak.

SECURITY

The Military Advisory Board,a panel of retired U. S. generals and admirals released a report entitled "National Security and the Threat of Climate Change. " The report predicts that global warming will have security implications, in particular serving as a "threat multiplier" in already volatile regions.

Britain's Foreign Secretary Margaret Beckett argues that "An unstable climate will exacerbate some of the core drivers of conflict, such as migratory pressures and competition for resources. " And several weeks earlier, U. S. Senators Chuck Hagel (R-NB) and Richard Durbin (D-IL) introduced a bill in the U. S. Congress that would require federal intelligence agencies to collaborate on a National Intelligence Estimate to evaluate the security challenges presented by climate change.

In November 2007, two Washington think tanks, the established Centre for Strategic and International Studies and the newer Centre for a New American Security, published a report analysing the worldwide security implications of three different global warming scenarios. The report considers three different scenarios, two over a roughly 30 year perspective and one covering the time up to 2100.

Its general results conclude that flooding ". . . has the potential to challenge regional and even national identities. Armed conflict between nations over resources, such as the Nile and its tributaries, is likely. . . " and that "Perhaps the most worrisome problems associated with rising temperatures and sea levels are from large-scale migrations of people — both inside nations and across existing national borders. "

Chapter 14

Resource Conservation with Modern Technique

Biotechnology has brought about a revolution in the way that plant genetic resources can be utilised. No longer are breeders confined to seeking genes useful for improvement among the closest relatives of a target genotype. Wide crossing can be facilitated by embryo rescue and potentially by protoplast fusion. Selective gene transfer by somatic cell genetic techniques ('genetic engineering') can, in theory, transcend the widest taxonomic barriers.

Related molecular technologies can aid disease indexing and germplasm characterization. At the more prosaic level, cloning *in-vitro* and the exchange of *in-vitro* plantlets make it possible to multiply and distribute elite genotypes with a speed, convenience and security far in excess of traditional technologies. Meristem-tip culture and shoot-tip grafting alone or combined with other forms of therapy can free plants of viral pathogens. This is particularly important for clonal crops that lack the 'filter' of seed production.

These advances in biotechnology, brought about by a combination of *in-vitro* (tissue culture) technology, genetics and molecular biology, are changing the way that we look at genepools and at the potential for crop improvement. The most dramatic benefits may well be reaped in the case of crops that are sterile and/or are traditionally propagated vegetatively.

Genetic engineering can bring about improvements in their hitherto virtually inaccessible genomes and the vegetative habit, amplified by *in-vitro* cloning, stabilizes the improved genotype.

Any resource of value clearly needs to be protected and plant genetic resources are no exception. Awareness of the value of genetic resources has been sharpened in recent years by threats to biodiversity in an often hostile world. This realization has come at the same time as the realization that new biotechnologies can greatly increase our ability to use genetic resources.

For crops that produce seeds capable of storage under conditions of low temperature and low humidity ('orthodox seeds') their genetic conservation can be managed very adequately by seed genebanks. However, this technical approach requires access to sufficient quantities of storable seeds. Significant amount of genetic resources fall outside that definition.

At one end of the spectrum are those materials that are not in cultivation or that have not been targetted for conservation efforts for other reasons. They are to be found in the natural environment and are probably best conserved there *'in-situ'*. At the other end of the spectrum are cultivated genotypes and their relatives that are recognised as being in need of *'ex-situ'* conservation but that do not respond to conventional seed storage methodology.

These problem crops fall into two categories: 'recalcitrant' seed producers and clonal crops. Recalcitrant seeds die when exposed to desiccating conditions or low temperatures. This presents serious problems to the breeder in both the short and long term. Seeds cannot be saved either from one generation of the crop to the next, nor can they be saved for future human generations. Examples of recalcitrant seed-producing species are mango, cacao, coconut, walnut, jackfruit and avocado.

Clonal crops cover a wide range of species from the root and tuber crops, such as potato, cassava, yam, taro and sweet potato to fruits, such as apple, pear, citrus, banana and the cooking banana (plantain). Other miscellaneous crops including vanilla, ginger, turmeric, hops and sugarcane are also clonally propagated.

In some of these cases, seed production is impossible due to sterility. In others, it is undesirable to produce seeds for conservation as this would break up highly heterozygous clonal genotypes. A significant number of the problem species are important in South and South-East Asia in particular, and the developing world, in general.

For both types of problem crops, conservation needs are met, traditionally by the use of the field genebank ('orchard', 'clonal repository', etc.). For reasons of security, cost, plant health and convenience, the field genebank has many disadvantages. I

ts advantages are few and relate to the opportunities that are offered for evaluation of the germplasm. However, this is not a function of a base genebank. In effect, recalcitrant seeds and clonal crops do not have, at present, a satisfactory, safe, base conservation mode.

In the rectification of this deficiency, we find another role for biotechnology in addition to those identified above. If genotypes can be inoculated into *in-vitro* culture for purposes of clonal propagation and germplasm distribution, why not prolong the *in-vitro* phase and use it for storage? The realization of the potential of *in-vitro* conservation came in the early 1970s at which time the storage of microbial cultures was virtually routine and the storage of animal tissue cultures a practical proposition, if not entirely routine.

In the last 20 years, we have seen dramatic advances in protoplast culture methodologies, in genetic transformation technology and in the extension of *in-vitro* propagation to most species that have been attempted. Over the same time scale, *in-vitro* conservation has moved from speculation to development and implementation. An explanation of the technologies involved in this evolution. However, it is relevant firstly to point out a further linkage between *in-vitro* conservation and biotechnology.

The different aspects of biotechnology involve the utilisation of a range of *in-vitro* culture systems: protoplasts, cells, callus, embryos, shoots and plantlets. Often, these are by their very nature ephemeral, highly dependent on external inputs for their continued health and growth and, in some cases, genetically unstable.

The phenomenon of 'somaclonal variation' is a double-edged sword that offers one way of generating variability *in-vitro* but that also threatens the stability of genotypes passing through the more susceptible unorganized culture systems. Accordingly, *in-vitro* cultures, the raw materials of plant biotechnology, have a conservation problem of their own. This has several aspects.

Firstly, there is a need to arrest time-related changes including somaclonal variation. Then there is a need in routine *in-vitro* culture work to store stock cultures in an economical and safe way, free of contamination and other risks. In experimental work, there is a need to 'suspend time' to enable direct comparisons to be made between samples processed over a period.

Finally, many of the genotypes evolved *in-vitro* do not have a natural storage form. Valuable but unstable callus cultures used in secondary product synthesis *in-vitro* may be incapable of conversion to a whole plant, having lost their 'totipotency'. Therefore, it follows that they cannot form seeds. In this way, the materials of plant technology themselves form a third problem category to add to the recalcitrant seed-producing and clonal crops.

In-vitro Conservation: Foundation Technologies

Although the essential core of plant genetic resources conservation is storage *per se*, the term 'conservation' covers the entire process from collecting through to access for evaluation and utilisation. This breadth of definition, encompassing a 'package' of technologies is true also for *in-vitro* conservation. The foundation technologies that make up an *in-vitro* conservation system are collecting, disease eradication and indexing, culture initiation, multiplication, storage and distribution.

As in conventional conservation, there are points where monitoring should take place to check viability and genetic stability as well as the important tasks of characterization and documentation.

Orthodox seed storage is carried out using methodologies that offer either 'active' or 'base' conservation conditions for the medium-term or long-term respectively and with different criteria in relation to access, study, multiplication, utilisation and distribution.

In parallel, *in-vitro* conservation has active and base components, which gives it an advantage over the field genebank which, as indicated earlier, is an active conservation mode. *in-vitro* active conservation is achieved by maintenance of cultures in the growing state, usually with growth slowed down by limiting physical environmental or culture medium factors.

In-vitro base conservation is achieved by cryopreservation, i. e. , storage in liquid nitrogen at or near a temperature of -196° C. The linkages between the different foundation technologies of *in-vitro* conservation and their relationship to the field genebank. It is clear that *in-vitro* conservation is more complex but, by the same token, has more to offer than conventional conservation.

With the exception of some examples in biotechnology involving, non-totipotent callus cultures as detailed earlier, culture initiation, multiplication and plant regeneration form the cycle of events that is the backbone of an *in-vitro* conservation scheme. The way in which this cycle is carried out has implications for the success of the conservation effort. There are two main areas where this is important.

The first is in relation to genetic stability. As already indicated, certain culture systems are less stable than others, being more prone to somaclonal variation. Therefore, where there is a choice, *in-vitro* conservation of plant genetic resources generally avoids the use of cell and callus cultures. (The use of protoplast cultures would be difficult to justify in any conservation context).

The remaining culture systems, embryos, shoots and plantlets are all more suitable. Until recently, there has been a strong prejudice favouring the use of shoots/shoot-tips/meristems for cryopreservation and shoots/plantlets for slow growth storage. While the latter choice can still be justified, there is an accumulation of evidence and experience that would say that embryo cultures have advantages in cryopreservation.

These largely relate to the second area of concern in the *in-vitro* culture cycle, i. e., amenability to storage. Structural factors, particularly size and degree of complexity, strongly influence the reaction of a culture system to cryopreservation. In general, the smaller and less complex the structure, the more likely it is to survive cryopreservation. This, unfortunately, is virtually the opposite of the criteria for judging the likelihood of somaclonal variation. A compromise may, however, be found in the somatic embryo (or indeed, the immature zygotic embryo) which may well combine a small unit size with a low risk of somaclonal variation.

The point of entry of germplasm into the *in-vitro* conservation cycle may vary. Sometimes, the germplasm will be transferred in the form of a culture from another *in-vitro* collection. Sometimes it will be inoculated *de novo* into culture in the *in-vitro* genebank. There is a third option, that of *in-vitro* collecting. This involves the introduction of material into culture actually in the field. Explores this novel and very useful yet simple application of an *in-vitro* technique in the service of genetic conservation.

In-vitro Collecting

Plant explorers have had many decades and more of experience in collecting germplasm from its natural environment or area of cultivation and transferring it to other locations and to genebanks. However, even the most experienced collector faces problems that threaten the efficiency of the collecting mission and the security of the germplasm collected. These problems waste resources and lead to inadequate sampling of target areas and genepools.

Even with good planning, it is possible to find insufficient supplies of seed or other suitable propagules or to find that the seed is immature. These problems may be due to seasonal variations or browsing by animals among other causes. In the case of vegetative propagules such as would be collected for say, banana, the propagule is prone to deterioration within a short period of time. This limits the duration of the collecting mission that can be undertaken and the area of territory that can be penetrated without complex transport arrangements for personnel and collected material.

Recalcitrant seeds are similarly prone to deterioration and have a second problem in some cases such as coconut, of exessive weight/ bulk. A crop such as cacao combines virtually all the problems, the two types of material normally collected being budwood that is prone to fungal attack in transit or large fruits containing recalcitrant seeds.

Over the last 8 years, IBPGR has been sponsoring research to develop an alternative, *in-vitro* based approach to collecting plant germplasm in the field.

This involves the use of minimal equipment, modified surface sterilization procedures to reduce toxicity and modified culture media, sometimes containing antibiotics and/or fungicides to control residual contaminants. Using these basic principles, methods have now been researched for collecting materials as diverse as cacao, coconut, forage grasses, temperate fruits and citrus.

In-vitro collecting is an approach rather than a rigid set of prescribed procedures. It needs and, indeed, offers to the collector great flexibility to adapt the precise methodology to his own situation. An example to illustrate the flexibility of the technique can be found in coconut.

Coconut embryo culture is well developed providing the foundation for a field collecting method that avoids the transportation of large quantities of heavy nuts. Several scientists have tackled the adaptation of embryo inoculation method using different approaches. In one case, considerable efforts have been made to reproduce the conditions of cleanliness found in the laboratory by using an inflatable glove box in which the standard sterlization, dissection and inoculation procedure can be carried out.

Rather less sophisticated is a second approach wherein plugs of endosperm containing the embryo are extracted in the field using a cork borer, surface sterilized and then dissected to reveal the embryos which are inoculated into culture. The only protection from contamination is afforded by a wind shield, a spirit lamp to flame instruments and necks of vessels and importantly, the dexterity of the scientist.

A further stage in the simplification of the procedure is to defer dissection and inoculation until the material reaches the laboratory. In this case, the sterilized endosperm plugs are inoculated into a sterile saline solution for transportation). The endosperm plugs are, if necessary, resterilized in the laboratory and dissected, the embryos then being transferred to normal medium.

Finally, we find a method that completely excludes any complex field procedure. The nuts are simply dehusked and cracked open in the field. The endosperm plugs are extracted and transferred to a plastic bag containing coconut water. The bag is closed and transferred to an insulated container for transport to a laboratory for sterilization, extraction and inoculation of the embryo in a laminar air-flow cabinet.

Efforts to test *in-vitro* collecting in a wide range of circumstances and with collector's new to in-*vitro* manipulations have been largely successful. Examples of field testing include collecting coconut in Indonesia and forage grasses in East Africa, in a collaboration between IBPGR and the International Livestock Centre for Africa. In 1990, IBPGR and CATIE in Costa Rica held a training course on *in-vitro* collecting supplementary collecting method alongside traditional approaches. It is a good example of appropriate biotechnology helping conservation.

In-vitro Storage: Principles, Successes and Continuing Challenges

There are two basic options for *in-vitro* storage, slow growth for the short-to-medium term and cryopreservation for the long-term (Since the definition of storage time-span and the concepts of active or base storage derive strictly from motivation rather than methodology, there is no reason why, in fact, cryopreservation should not have applications in short-to-medium term conservation.

The principle of slow growth storage is that the safety of *in-vitro* culture be used without disadvantages of frequent subculturing. Thus, the inputs in terms of labour and consumables are reduced, as also the risks of contamination at each transfer interval. However, the cultures can be observed to be growing and can readily be restored to normal culture conditions for multiplication and distribution to users.

The subculturing interval can, in a few fortunate cases, be extended without significant modification of culture procedures. Generally, it requires culturing at a reduced temperature or in the presence of retardants.

The disadvantages of slow growth are few but relatively important. The stress imposed on a culture to retard its growth will, inevitably, affect its vitality. A careful balance needs to be maintained to avoid general loss of viability and, worse, selective loss of viability that could lead to genetic drift. This might seem to be a strange concept in work with putative clones.

However, it is probably safer to consider *in-vitro* collections as 'tight populations' rather than absolute clones. This is certainly the case for unorganized cultures, such as callus, and may be true for some organized material. For these reasons, whilst slow growth storage does reduce the direct inputs to culture maintenance, it still gives a burden of monitoring efforts to the curators of the culture collection.

Despite these cautions, we can find an impressive number of notable successes in the routine maintenance of germplasm in the form of shoot cultures or plantlets in slow growth. Significant examples can be found within the IARCs for root and tuber crop germplasm, the International Potato Centre (CIP) maintaining collections of potato and sweet potato, the Centro Internacional de Agricultura Tropical (CIAT), cassava; and the International Institute of Tropical Agriculture (IITA), yam and cassava.

Other examples include the banana and plantain collection maintained at the INIBAP Transit Centre in Leuven, Belgium, and numerous collections of clonal crops conserved by national programmes and research laboratories around the world.

The continuing challenges with slow growth storage fall into four categories. Firstly, we do not know the breadth of applicability of the technique in terms of genotypes within a species to which it is suitable or among species.

As with the development of *in-vitro* propagation technology, it will, for the most part, be a case of trial and error with only limited scope for anticipating reactions and needs of particular genotypes. This will all take time and individual efforts on the part of those wishing to apply the technique.

Secondly, there has not in any one case been an absolutely full evaluation of the interactions of different environmental factors (basic medium composition; retardants, temperature, light level, quality and photoperiod). Therefore, we know little about the latitude for varying one or more of these factors. This makes technology transfer potentially problematic, since not all the factors are documented even if appreciated.

Thirdly, we do not know the maximum period of time for which slow growth can safely be applied. There is a reasonable amount of information available on the duration of the typical subculturing interval but not on how many can be applied sequentially or whether the interval remains constant over time. There are anecdotal suggestions of cultures adapting ('habituating') to slow growth conditions.

Finally, there is a knowledge gap in relation to genetic stability or even phenotypic stability in material exposed to slow growth conditions for extended periods of time. There is a limited amount of information available from studies of callus cultures to suggest that there may either be selection or persistent physiological change brought about by slow growth at reduced temperatures.

The same observation has recently been made for callus stored under mineral oil overlay. Evidence on organized cultures is lacking. Accepting that we should expect them to be more stable, it is still necessary to apply a degree of caution in recommending their storage in slow growth without reservation for extended periods of time.

In summary, the challenges with slow growth storage are to press on with applying it as widely as possible, learning about its scope and limitations, and gathering data to build up a picture of its effects upon cultures in all respects.

Cryopreservation is a very different phenomenon from slow growth with its own advantages, disadvantages and challenges. Because all metabolism is suspended at the temperature of liquid nitrogen (- 196° C) or, as is often used for storage, liquid nitrogen vapour (- 150° C), time is not a critical factor in the design and application of cryopreservation regimes.

Provided that stable storage conditions be maintained, the only known cause of damage during cryopreservation is free radical activity. This may be initiated by endogenous reactions or may be initiated by background radiation. In either case, some protection can be afforded by chemicals such as dimethyl sulphoxide (DMSO) which are 'radioprotectant' as well as cryoprotectant.

Rather than time in storage being an important factor in cryopreservation, the most critical stages are the transition to and from the frozen state. With the exceptions of some extremely cold tolerant/ freeze tolerant materials, this is a highly unnatural transition and measures must be taken to protect both cellular and tissue structure and the metabolic processes against deleterious effects. These measures can and need be applied at each stage of cryopreservation process: pregrowth, cryoprotection, cooling, warming, the immediate post-thaw period and recovery growth.

There is now great accumulation of information on the variables applying to each stage. In the case of cell suspension cultures, a virtually routine cryopreservation method is available (Withers and King, 1980). For other systems, this is not the case.

A summary of the various methodologies attempted is not easy to make. There are great differences between culture systems and also interactions within and between the various stages. Thus, a particular cryoprotectant or recovery medium may be more effective at one cooling rate than another. However, some general observations can be made concerning the progress that has been made in elucidating different stages and what remains to be researched.

'Pregrowth' refers to the period of time (usually measured in days) in culture before cryopreservation proper begins. It offers opportunities to induce physiological changes in the specimen to increase its freeze-tolerance. Most effective pregrowth treatments involve some type of 'stress conditioning' (high osmoticum, low temperature etc.).

Among the most interesting observations made in recent years regarding pregrowth are the finding that the use of high levels of sucrose (0. 75M) for somatic embryos of oilpalm can eliminate the need for cryoprotectants and reduce dependence on precisely 'tuned' cooling rates and that cold hardening *in-vitro* combined with controlled desiccation can mimic the *in-vitro* hardening of fruit buds enabling the cultures to be frozen, again without cryoprotectants.

Cryoprotection is the phase of minutes to hours during which single protectant compounds or several-component mixtures are applied. The most commonly used cryoprotectant is DMSO which can also act as an effective pregrowth additive. DMSO can be used alone for organized cultures. It is more frequently used in mixtures, e. g. with glycerol and sucrose for unorganized cultures.

Sucrose is a naturally occurring compound in plants arid a little research has been carried out in the use of compounds such as proline that are known to be induced under conditions of stress in nature. However, in general, there would appear to be great scope for exploring the potential of natural cryoprotectants, either induced by pregrowth stress conditions such as outlined above or exogenously applied. One of the attractions of using more natural compounds is that it reduces the risk of cryoprotectant toxicity, a problem that can be encountered with high concentrations of DMSO and with old or impure stocks of the chemical.

Cryoprotection is followed by cooling. This can be carried out in a number of ways depending upon the mechanism of survival in use. The mechanism also has implications for the concentrations and types of cryoprotectants used. Two basic mechanisms exist. The first and most commonly used is that of protective dehydration. Slow cooling will cause extracellular ice formation which leads to the extraction of water from the cell and a reduction in the amount of potentially damaging ice formation when the cell contents eventually solidify.

There can be a fine balance between under and over dehydration. It is important to find this as overdehydration can lead to 'solute damage' and excessive shrinkage of the cell with effects such as the loss of membrane surface area causing membrane rupture upon re-expansion. There are experimental procedures to arrive at the optimum amount of dehydration.

In some slow cooling regimes, it is beneficial to remove the superficial water surrounding the specimen by blotting using a filter paper. It is even possible to carry out a certain amount of dehydration in an air stream or in a desiccator before cooling.

The alternative to slow cooling is rapid cooling which avoids ice damage through the induction of either microscopically small ice crystals or avoids ice formation totally through the process of 'vitrification'. Expedients such as direct exposure to liquid nitrogen have been used to achieve survival in specimens treated with conventional, relatively low concentrations of cryoprotectants.

However, vitrification uses extremely high concentrations that are potentially toxic at non-frozen temperatures. Although they facilitate beneficial solidification of the specimen into an ice-free mass, they can cause damage between the time of addition and removal. Therefore, a lot of attention is given in vitrification work to the processes termed 'loading' and 'unloading'.

Vitrification has proved highly successful for animal embryos. It has been applied to various *in-vitro* culture systems from protoplasts to shoot-tips. In the latter case, there is the greatest potential benefit since, as indicated below, large and highly organized specimens tend to suffer structural injury in dehydration-based cryopreservation procedures.

Warming to achieve thawing is simply carried out but is a critical stage in the cryopreservation process. Where there is a risk of ice recrystallization as in rapidly cooled specimens, it is essential that thawing be carried out rapidly as well. There are some examples of successful, even beneficial slow thawing after slow cooling but, in general, it is wise to use rapid thawing as a first option.

The freshly thawed specimen is extremely vulnerable. In the literature, we can find examples of sensitivity to rehydration and plasmolysis injury manifested as weakness in the cell membranes. There is evidence for the infliction of metabolic lesions including respiratory stress. Cell division is delayed and structural injury is visible at gross and microscopical/electron microscopical levels, with corroborative biochemical data indicating membrane breakdown.

Sensitivity to high light levels is detected in some cases, probably indicating photoxidation involving free-radical mechanisms. The frozen and thawed cell is likely to have suffered metabolic insults that lead to the disruption of the normally efficient free-radical scavenging mechanisms in the cell.

There is some evidence that the provision of free radical scavengers can be beneficial during the early stages of recovery. In general, it is wise to take all measures to stabilize the specimen at this stage by incubating under low light levels, by avoiding osmotic disturbance (e. g. by washing) and by using semi-solid rather than liquid medium. If it is felt necessary to remove residual cryoprotectants, this can be done by draining on a filter paper and/or moving the specimen, carried on a filter paper, through a series of dishes of fresh medium to allow the cryoprotectant to diffuse away.

The recovery stage involves the resumption of normal metabolic processes, the resumption of cell division and the further growth and/or development of the specimen. In the case of callus cultures and cell suspensions, this is a relatively simple process involving the accumulation of a mass of undifferentiated cells.

In the case of organized cultures, it is more complex and involves the re-initiation of meristematic activity. If the specimen has been severely damaged in freezing and thawing, this may be problematic. It is a common observation that shoot-tips undergo a period of undifferentiated growth through callus formation either alongside development of the original shoot meristem or in place of it.

In the latter situation, there is a serious risk of somaclonal variation, in the former case the risk is less but not to be ignored. The use of growth regulators to stimulate growth after freezing and thawing can exacerbate this problem and the use of auxins in particular is best avoided. The problems of significant damage in both quantitative and qualitative terms encountered in the cryopreservation of organized structures is probably the main motivation for exploring the use of vitrification. However, the scope for modifying the conventional slow cooling-based cryopreservation method should not be underestimated. Systematic experimentation to optimise the variables at the pregrowth, cryoprotection, cooling and recovery stages can yield great benefits.

Vitrification is one of the two most interesting new developments of recent years in cryopreservation; the other is the adaptation of artificial seed technology. Artificial seeds have been explored as a means of supporting somatic embryos, both nutritionally and structurally, of transporting and delivering them to give a degree of convenience approaching that of natural or pelletted seeds. Success in these endeavours has been significant, although there is much research to do to make the technology universally applicable and economic.

One of the exciting spin-offs of this research may be an improvement in cryopreservation technology for organized cultures such as somatic embryos and shoot-tips, taking advantage of research into the desiccation process as well as the encapsulation itself.

There is already evidence that the encapsulation of shoot-tips is a practical approach to cryopreservation that may lead to improved survival and maintenance of organization. This is an area of research to follow closely, particularly as it can take advantage of the attractive somatic embryogenesis propagation system.

In-vitro Conservation of key Crops

An evaluation of the current status of *in-vitro* conservation can be gained by reviewing the progress that has been achieved in creating the 'package' of technologies that complete an *in-vitro* conservation scheme for a given crop. The target crops for *in-vitro* conservation are many and diverse. Three instructive examples to review in this way are coconut, banana/plantain and cassava. They are all important staple crops in developing countries and have all received a significant amount of attention from the point of view of *in-vitro* culture.

In the context of *in-vitro* conservation, coconut is important as a recalcitrant seed-producing crop. It has received a lot of attention from *in-vitro* culture technologists because it is impossible to clone by conventional means. The clonal propagation *in-vitro* of oilpalm has become routine in recent years.

Despite some concerns over genetic stability in certain propagation systems, the advantages of being able to clone oilpalm are tremendous and this has served as a model to encourage efforts on coconut. When we look at the state of the package of *in-vitro* conservation technologies for coconut, however, these efforts are not yet reflected in successes.

In-vitro collecting is highly developed and the same technology can be used to exchange germplasm between genebanks. Unfortunately, though the culture system used involves one-to-one regeneration from zygotic embryos, this severely limits the realization of the potential of both *in-vitro* cloning and *in-vitro* storage.

Technically, cultured plantlets germinated from zygotic embryos could be subjected to slow growth storage but the only advantage that this would offer over the field genebank would be potential protection from infection with pathogens. Without extensive investigation to optimize slow growth conditions for a sufficiently wide range of genotypes, it would be an extremely risky conservation mode since it lacks the opportunity to replicate valuable genotypes in culture.

Cryopreservation of mature zygotic embryos has been achieved with some very limited success. Early indications from IBPGR-sponsored research carried out by ORSTOM/IRHO in France and Cote d'Ivoire are that immature embryos of coconut may be responsive to Cryopreservation. However, the problem remains of applying storage to unreplicated culture units.

The lack of a clonal propagation technique is an extremely serious bottleneck in the conservation and utilisation of coconut germplasm by biotechnological means. Other lacunae in the case of coconut exist in the area of disease indexing, disease eradication and quarantine. There may be scope for some *in-vitro* approaches here, particularly in providing a contained quarantine method.

The biotechnological modification of coconut genotypes is virtually unexplored and will again meet a bottleneck in the area of clonal propagation.

Bananas and plantains present a different type of challenge from coconuts. Some but not all genotypes produce seeds. The storage of seeds has been found difficult but they are not recalcitrant.

Vegetative propagation is the normal method of multiplication and clones are the main target for conservation efforts. In respect of *in-vitro* techniques, a much more satisfactory situation exists for banana and plantain, where clonal propagation is virtually routine, *in-vitro* collecting has been attempted with some success but not widely tested, and *in-vitro* exchange and slow growth are used in routine.

The INIBAP Transit Centre in Leuven, Belgium among others exchanges cultures and maintains a large collection in slow growth. Meristem-tip culture has had some success in eliminating viruses from shoot-tips in this crop but is not entirely satisfactory due to the large size of the meristem compared with other examples such as potato and cassava.

Success in attempts to cryopreserve embryogenic suspension cultures of a cooking banana cultivar (Bluggoe) has been reported recently. The Cryopreservation method is based on cryoprotection with DMSO and slow cooling. Plant regeneration has been achieved through somatic embryogenesis. Earlier attempts to cryopreserve the large and fleshy shoot-tip of banana were totally unsuccessful, illustrating the advantages that can be gained by the use of embryogenic systems.

However, a major problem with bananas and plantains relates to genetic stability *in-vitro.* Some genotypes in particular appear to be prone to high levels of instability that seriously affect field performance. There is a need to determine the causes of this instability to understand better its linkage to genotype, to develop methods for detecting instability at a sufficiently early stage in the propagation/storage cycle, and to determine whether storage conditions affect the incidence of instability either positively or negatively.

The improvement of banana and plantain genotypes by using biotechnological approaches must be considered within reach. The range of culture systems available in the genus is sufficient to support genetic manipulation studies and there are already reports of the successful isolation of somaclonal variants with fusarium wilt resistance. The disease aspects of coconut are equally applicable to banana and plantain.

Let us look now at the final model chosen for this analysis. Cassava is normally propagated vegetatively and most conservation efforts are devoted to the storage of clones, despite the fact that seeds can be produced and stored. Cassava has been field collected using an *in vitro* approach (Roca, personal communication). It is readily introduced into culture and propagated clonally via shoot nodal cuttings and can be induced to form somatic embryos. Slow growth storage is carried out in routine. IBPGR and CIAT are currently finalizing a project to test management standards for *in-vitro* active genebanks using cassava as a model.

Cryopreservation of cassava seeds and zygotic embryos is feasible and several scientists have investigated the possibility of cryopreserving *in-vitro* cultures. It is now possible to achieve survival and organised recovery at a high frequency in the case of shoot-tips, and somatic embryos are also found responsive.

There is now the foundation for developing an *in-vitro* base genebank for cassava. *In-vitro* exchange is routine for this crop and, in combination with meristem-tip culture and associated techniques to eradicate virus, makes a major contribution to the health status of cassava germplasm in circulation and conservation.

Efforts are currently under way to apply genetic manipulation techniques to cassava with a view, in the first instance, to transferring pest and disease resistance. Therefore, of the three models, cassava is having the closest fully feasible and operational *in-vitro* conservation and utilisation package.

While comparison of the situation for cassava with that of coconut might give cause for disappointment in the case of the latter crop, it should be remembered that it is only very recently that the picture was so optimistic for cassava.

Once the major bottleneck of *in-vitro* clonal propagation has been solved for coconut, the path will be clear for developing the other components of the package.

Complementary Conservation Strategies

Rather than concentrating thoughts on technologies, the starting point for devising a conservation strategy should be the genepool. The needs of a genepool are very unlikely to be met by a single conservation technology.

While it is not always the case in practice, it is eminently logical to have a balanced application of both *in-situ* and *ex-situ* approaches to conservation. Within the latter category, a further balance needs to be struck between seed, field genebank, *in-vitro,* pollen and, perhaps in the future, DNA storage.

Each technology should be chosen on the basis of utility, security and complementarity to the other components of strategy. The inappropriate application of sophisticated technologies, such as the unnecessary use of *in-vitro* cloning and storage, or its use without a sufficiently rigorous experimental foundation is highly inadvisable. It may endanger germplasm through storing it under suboptimal conditions or with an inadequately wide genetic base, it will damage the reputation of biotechnology, and will certainly waste financial resources.

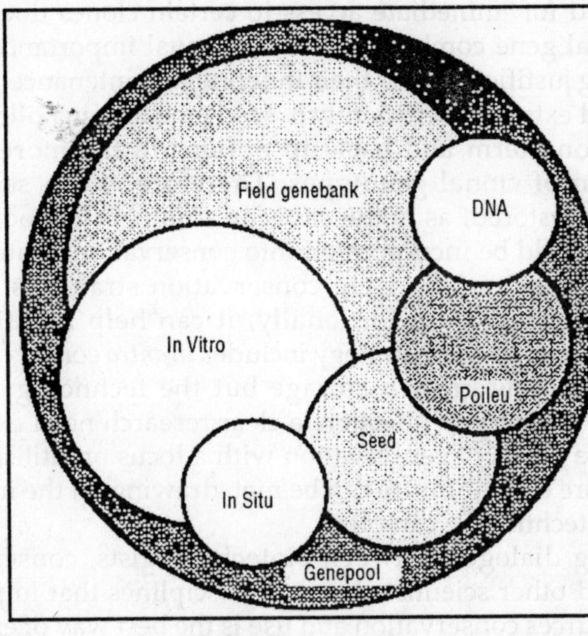

Fig. A Hypothetical Model Showing the Complementary Use of Different Conservation Technologies

This caution applies particulary to the use of molecular technologies. They have a number of very valid applications in the improvement of genotypes by selective gene transfer, in increasing our knowledge of genetic diversity by the application of Restriction Fragment Length Polymorphism (RFLP) and other associated technologies alongside isozyme analysis and conventional morphological characterization, and improvement of disease diagnostics by the development of broad and narrow spectrum probes.

However, whilst some specific niches can be foreseen for the use of DNA storage within a complementary conservation strategy, it is difficult to see this approach safely overtaking the other available conservation technologies.

For the problem crops considered here, a significant part of the genepool will certainly be conserved either in the field genebank or *in-vitro* for the foreseeable future. The two approaches are truly complementary, each balancing the advantages and disadvantages of the other. The security of *in-vitro* conservation and independence of climatic conditions balance the risks and climatic specificity of the field genebank. *In-vitro* conservation is also likely to be less expensive. However, the field genebank is an important adjunct to improvement programmes since it allows evaluation.

The need for immediate access to certain clones does not mean that all clonal gene combinations are of equal importance over time with a strong justification for their indefinite maintenance. Genes may be conserved extremely efficiently in seed (and also in pollen), offering important long-term and back-up options to the more accessible conservation of clonal genotypes. Therefore, when seeds can be produced and stored, as is the case for a number of root and tuber crops, this should be incorporated into conservation strategies.

This approach of devising conservation strategies should aid efficient conservation. Additionally, it can help identification of research needs. If an ideal strategy includes *in-vitro* conservation under conditions of slow growth storage but the technology is not yet available for the crop in question, a clear research need exists. In this way, the true needs of conservation with a focus on utilisation in the near and more distant future can be met, drawing on the inputs of the best that biotechnology can offer.

A strong dialogue between biotechnologists, conservationists, breeders and other scientists in many disciplines that impinge upon genetic resources conservation and use is the best way of ensuring the necessary stimulus to research, its adaptation to pressing problems and the realization of new bio-technological solutions in agriculture.

Efforts to develop conservation strategies that incorporate biotechnological approaches have been made by the International Agricultural Centres of the Consultative Group on International Agricultural Research (CGIAR) along with their partners in National Programmes. Some examples of the contributions already made to the overall global effort by the IARCs.

IBPGR collaborates with many partners in such efforts. In research itself, IBPGR makes efforts through international crop genetic resources networks to bring together users and curators of crop germplasm. They, collectively with scientists doing research on the technologies necessary to store, document and utilise the germplasm, and have the necessary expertise to develop priorities for action to enhance the safe conservation and utilisation of crop genepools.

Biotechnology has applications in the utilisation of plant genetic resources through mass clonal propagation *in-vitro*, in genetic manipulation, in aspects of phytopathology and safe distribution of germplasm, in characterization and, importantly, in conserving germplasm *in-vitro* to reinforce the benefits of the other applications. *In-vitro* conservation is particularly valuable for crops that produce recalcitrant seeds and those that are clonally propagated.

Their conventional conservation in the field genebank is hazardous in several respects. *In-vitro* conservation is a package of technologies commencing with collecting and leading through disease indexing, disease eradication and storage eventually to distribution and utilisation. *In-vitro* collecting is useful in overcoming problems relating to the availability, condition and portability of germplasm to be gathered in the field. *In-vitro* storage involves the establishment of cultures, preferably of a type that carry a minimal risk of somaclonal variation, and their transfer to slow growth or cryopreservation.

Slow growth is usually achieved by reducing the culture temperature or adding growth retardants to the culture medium. *In-vitro* base storage can only be achieved by cryopreservation, that is, storage in liquid nitrogen. All types of culture have been successfully crypreserved but cell cultures are the most easy to handle and shoot-tips the most problematic.

A new approach to cryopreservation termed 'vitrification' offers the potential" to overcome problems experienced with highly organized material such as shoot-tips. Another exciting new departure is the involvement of artificial seed technology in conservation work. The stage of development of *in-vitro* conservation varies from crop to crop with cassava being the most highly developed of the problem crops for which this approach is beneficial, and coconut being among the least well developed.

In these as in other cases, *in-vitro* conservation will never replace conventional technologies entirely but will complement them within a strategy that balances of the traditional approaches of *in-situ*, seed and field genebank conservation with the best of the new using where appropriate, *in-vitro* conservation, pollen storage and even DNA storage.

Index